Lecture Notes in Mechanical Engineering

Lecture Notes in Mechanical Engineering (LNME) publishes the latest developments in Mechanical Engineering—quickly, informally and with high quality. Original research reported in proceedings and post-proceedings represents the core of LNME. Volumes published in LNME embrace all aspects, subfields and new challenges of mechanical engineering. Topics in the series include:

- Engineering Design
- Mechanical Structures and Stress Analysis
- Engine Technology
- Nanotechnology and Microengineering
- Control, Robotics, Mechatronics
- MEMS
- Theoretical and Applied Mechanics
- Fluid Mechanics
- Engineering Thermodynamics, Heat and Mass Transfer
- Manufacturing
- Precision Engineering, Instrumentation, Measurement
- Tribology and Surface Technology

To submit a proposal or request further information, please contact the Springer Editor of your location:

Europe, USA, Africa: Leontina Di Cecco at Leontina.dicecco@springer.com
China: Ella Zhang at ella.zhang@springer.com
India: Priya Vyas at priya.vyas@springer.com
Rest of Asia, Australia, New Zealand: Swati Meherishi
at swati.meherishi@springer.com

Indexed by SCOPUS and EI Compendex. All books published in the series are submitted for consideration in Web of Science.

To submit a proposal for a monograph, please check our Springer Tracts in Mechanical Engineering at https://link.springer.com/bookseries/11693

Magd Abdel Wahab
Editor

Proceedings of the 10th International Conference on Fracture Fatigue and Wear

FFW 2022, 2–3 August, Ghent University, Belgium

Springer

Editor
Magd Abdel Wahab
Laboratory Soete, Faculty of Engineering
and Architecture
Ghent University
Zwijnaarde, Belgium

ISSN 2195-4356 ISSN 2195-4364 (electronic)
Lecture Notes in Mechanical Engineering
ISBN 978-981-19-7807-4 ISBN 978-981-19-7808-1 (eBook)
https://doi.org/10.1007/978-981-19-7808-1

This Springer imprint is published by the registered company Springer Nature Singapore Pte Ltd.
The registered company address is: 152 Beach Road, #21-01/04 Gateway East, Singapore 189721, Singapore

Organising Committee

Chairman

Prof. Magd Abdel Wahab, Ghent University, Belgium

International Scientific Committee

Prof. S. Abdullah, Universiti Kebangsaan Malaysia, Malaysia
Dr. J. Abenojar, Universidad Carlos III de Madrid, Spain
Dr. A. H. Ertas, Bursa Technical University, Turkey
Prof. T. Hattori, Shizuoka Institute of Science and Technology, Japan
Dr. M. Kchaou, University of Sfax, Tunisia
Dr. S. Khatir, Ghent University, Belgium
Dr. C. Le Thanh, Open University Ho Chi Minh City, Vietnam
Dr. K. Masuda, University of Toyama, Japan
Prof. T. Miyazaki, University of Ryukyu, Japan
Prof. H. Nguyen, HUTECH, Vietnam
Prof. K. Oda, Oita University, Japan
Prof. R. V. Prakash, Indian Institute of Technology, India
Prof. T. Qin, China Agricultural University, China
Prof. F. Rezai-Aria, Ecole des Mines d' Albi, France
Prof. T. Rabczuk, Bauhaus University Weimar, Germany
Dr. A. Rudawska, Lublin University of Technology, Poland
Dr. M. S. Prabhudev, Government Polytechnic, Kalgi, India
Prof. J. Song, Ostwestfalen-Lippe University, Germany
Prof. J. Toribio, University of Salamanca, Spain
Prof. C. Xu, China Agricultural University, China

Dr. T. Yue, Ghent University, Belgium
Prof. C. Zhou, Nanjing University of Aeronautics and Astronautics, China
Dr. X. Zhuang, Leibniz Universität Hannover, Germany

Preface

This volume contains the proceedings of the 10th International Conference on Fracture Fatigue and Wear FFW 2022, which is held online during the period 2–3 August 2022. Previous conferences were celebrated in Jinan (China, 2010), Kitakyushu (Japan, 2013), Kitakyushu (Japan, 2014), Gent (Belgium, 2015), Kitakyushu (Japan, 2016), Porto (Portugal, 2017) and Gent (Belgium, 2018–2021).

The overall objective of the conference is to bring together international scientists and engineers in academia and industry in fields related to fracture mechanics, fatigue of materials, tribology and wear of materials. The conference covers industrial engineering applications of the above topics including theoretical and analytical methods, numerical simulations and experimental techniques. One of the aims of the conference is to promote cooperation between international scientists and engineers from a large number of disciplines, who are involved in research related to fracture, fatigue and wear. The presentations of FFW 2022 are divided into three main sessions, namely (1) fracture, (2) fatigue and (3) wear.

The organising committee is grateful to the keynote speaker, Prof. Dr. Jesús Toribio, Fracture and Structural Integrity Research Group (FSIRG), University of Salamanca (USAL), Spain, for his interesting keynote speech entitled 'Innovative Approach to Fatigue Crack Propagation on the Basis of Microstructurally-Induced Locally-Deflected Crack Paths: A Tribute to Antonio Machado and Fray Luis de León', and to Keynote Speaker Prof. Luca Susmel, the University of Sheffield, UK, for his interesting keynote speech entitled 'Cracking behaviour and static assessment of notched additively manufactured polylactide'.

Special thanks go to members of the Scientific Committee of FFW 2022 for reviewing the articles published in this volume and for judging their scientific merits. Based on the comments of reviewers and the scientific merits of the submitted manuscripts, the articles were accepted for publication in the conference proceedings and for presentation at the conference venue. The accepted papers are of a very high scientific quality and contribute to the advancement of knowledge in all research topics relevant to FFW conference.

Finally, the organising committee would like to thank all the authors, who have contributed to this volume and to those who have presented their research work at the conference venue.

Zwijnaarde, Belgium

Prof. Magd Abdel Wahab
Chairman of FFW 2022

Contents

Fracture

A Few Fracture Features of Al-Based and Cu-Based Ribbon Metallic Glasses Under Non-isothermal and Oscillating Loading 3
Arseniy Berezner, Victor Fedorov, and Gregory Grigoriev

Finite Element Failure Analysis of Single-Lap Bolted Connection Under Impact Load 11
Mingpo Zheng, Zhifeng Liu, and Magd Abdel Wahab

Plasticity Analysis in Aluminum Alloy Plates Repaired with Bonded Composite Patch Under Overload 21
Faraz Ahmed, S. M. A. K. Mohammed, Faycal Benyahia, Bel Abbes Bachir Bouiadjra, and Abdulmohsen Albedah

Machine Learning for Predicting Pipeline Displacements Based on Soil Rigidity 29
Meriem Seguini, Samir Khatir, Djamel Nedjar, and Magd Abdel Wahab

Delamination Detection of Rectangular Laminated Composite Plates by Combining the One-Dimensional and Two-Dimensional Discrete Wavelet Transforms 41
Morteza Saadatmorad, Ramazan-Ali Jafari-Talookolaei, Mohammad-Hadi Pashaei, Samir Khatir, and Magd Abdel Wahab

A Feasibility Review of Novel Avian-Based Optimization Algorithms for Damage Detection in a Truss Bridge 53
Lan Ngoc-Nguyen, Samir Khatir, Hoa Ngoc-Tran, Huu-Quyet Nguyen, Long Ngoc-Nguyen, Thanh Bui-Tien, and Magd Abdel Wahab

Static Assessment of Notched Additively Manufactured Polymers Based on the Theory of Critical Distances 65
Luca Susmel

Fatigue

Modeling Crack Initiation in Low Cycle Fatigue: A Review 79
Hussein Zahran, Aleksandr Zinovev, Dmitry Terentyev, and Magd Abdel Wahab

Investigating the Effect of the Characteristics of Inclusions and Micro-voids on Fretting Fatigue Initiation Lifetime with Two-Scale Analysis Approach 91
Can Wang and Magd Abdel Wahab

Wear

Effects of Load and SiC_p Addition on Wear Behaviour of Powder Injection Moulded Aluminium Composite 105
Tapany Patcharawit, Arada Ngeekoh, and Nutthita Chuankrerkkul

About the Editor

Prof. Magd Abdel Wahab is Full Professor of Applied Mechanics in the Faculty of Engineering and Architecture at Ghent University, Belgium. He received his B.Sc., 1988, in Civil Engineering and his M.Sc., 1991, in Structural Mechanics, both from Cairo University. He completed his Ph.D. in Fracture Mechanics in 1995 at KU Leuven, Belgium. He was awarded the degree of Doctor of Science from the University of Surrey in 2008. He has published more than 600 scientific papers in solid mechanics and dynamics of structures and edited more than 30 books and proceedings. His research interests include fracture mechanics, damage mechanics, fatigue of materials, durability, and dynamics and vibration of structures.

About the Editor

Fracture

A Few Fracture Features of Al-Based and Cu-Based Ribbon Metallic Glasses Under Non-isothermal and Oscillating Loading

Arseniy Berezner, Victor Fedorov, and Gregory Grigoriev

Abstract In the work, some analytical relationships on deformation of Al-based and Cu-based metallic glasses under variable heating and oscillating mechanical loading are given. It was shown that using of hydrodynamic laws with a nonlinear deformation model can be possible for fracture of ribbon metallic glasses. Behaviour of linear thermal expansion coefficient is analysed, and its role in the plastic deformation of metallic glass is estimated. Obtained results agree with experiments, and they can be used both in fundamental and applied investigations.

Keywords Metallic glasses · Thermo-mechanical processes · Plastic deformation · Microscopy

1 Introduction

It is known [1] that under the different dissipative factors, a system can become probably by deformation into a new stable state. Herewith, a fixed set of external conditions can lead to a personal deformation regime [2]. For example, creep [3, 4], dynamic (or thermal) mechanical analysis (DMA or TMA) [5] and others are that processes. For their model description, besides continuum mechanics (either the integral values or the finite sums), a structural (molecular or atomic) approach, based on the specific system defects and their features [6], is used. However, modelling of the plastic and fracture stages in frames of continual mechanics is quite difficult because all relationships clearly deviate from the linear laws such as Kelvin–Voigt model and so on. Thus, additional approaches, accounting generation and growth the critical system defects (dislocations, pores or cracks and others), are proposed for a solving [7]. In some cases, hydrodynamic and thermodynamic methods are used. At sufficient deviation of a specimen from linear geometry by a form (corrugation, necking or turbulent mass transfer), modelling is complicated by necessity in the special distribution of scalar or vector fields, which could describe real situation.

A. Berezner (✉) · V. Fedorov · G. Grigoriev
Derzhavin Tambov State University, Internatsionalnaya Str. 33, Tambov 392000, Russia
e-mail: a.berezner1009@gmail.com

M. Abdel Wahab (ed.), *Proceedings of the 10th International Conference on Fracture Fatigue and Wear*, Lecture Notes in Mechanical Engineering,
https://doi.org/10.1007/978-981-19-7808-1_1

Numerical modelling [8] (such as finite element method, etc.) is one of the alternative ways, but it not always gives the simple applied formulas. And description by an empirical or a hypothetical finite system of equations can be a more effective method for solving this problem.

Shear bands [9] and their zones (STZ) [10], whose redistribution leads to deformation of the specimens, are often proposed as the main defects of amorphous alloys (i.e. metallic glasses, MGs). But the complicated deformation dynamics and lack of the exhaustive models for MGs do not permit a full description of plastic flow and fracture under variable temperatures combined with mechanical loading. Because of that, the goal of this work is derivation of some model relations, describing the deformation and fracture features of ribbon amorphous alloys, which undergo heating and periodical mechanical loading.

2 Materials and Methods

As a basement of the modelling, our previous experimental data on DMA of ribbon Al-based and Cu-based MGs [11, 12] were chosen. Fractography of the specimens was observed by scanning electron microscopy (SEM, Jeol JCM-7000). The general factors for all DMA were preloading F_{load} (constant during the whole experiment) with additional one A (oscillated with $\omega = 6\pi$ rad/s frequency), which acted on the specimens. And continuous heating of constant rate $V_T = 5$ K/min with time t up to the fracture moment B of a specimen was carried out. During every experiment, the specimen elongated from l_0 ~18.5 mm initial size to critical length ~27 mm. Schematic plots of the whole process are presented in Fig. 1 for reaction force F and deformation ε.

As seen from the figure, observed curves deviate from Kelvin–Voigt typical behaviour that causes additional analysis of the experimental conditions. In this case, our previous equations for variable l deformation and reaction force $F_{react.}$ of the material [13] must be mentioned here:

$$l(t) = l_0 + \frac{Ct}{B^2 - Bt} \tag{1}$$

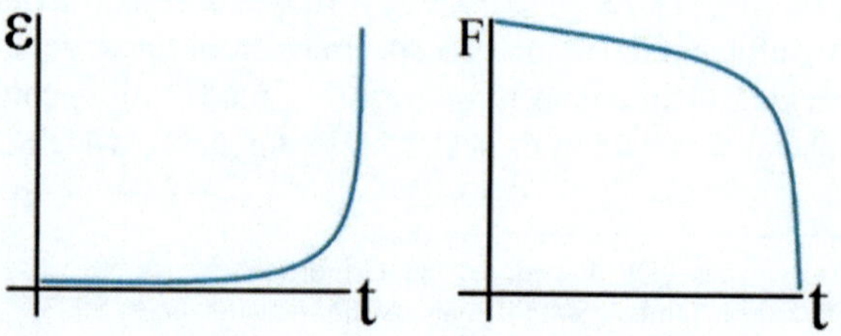

Fig. 1 Deformation dynamics of a specimen, occurring at heating with oscillated load, which is set except the static one

Table 1 Personal material parameters and experimental conditions

	C [m·s]	B [s]	m [kg]	F_{load} [N]	A [N]
$Al_{85}Y_8Ni_5Co_2$	0.078	3157	10^{-4}	3.6	0.0036
$Cu_{54}Pd_{28}P_{18}$	0.0492	2827	10^{-4}	0.9	0.0063

$$F_{\text{react}}(\omega, t(T)) = F_{\text{load}} - A\sin(\omega t) - \frac{2mC}{(B-t)^3}. \tag{2}$$

With respect to uniform temporal heating from the initial temperature T_0 (300 K) to variable one T (not above the crystallization point T_x: 550 K for Al-based and 568 K for Cu-based alloys), we can describe the heating regime as:

$$t = \frac{T - T_0}{V_T} = \frac{\Delta T}{V_T}. \tag{3}$$

For Al-based and Cu-based alloys, there are typical mean values that are given in Table 1, where C is the personal deformation coefficient of material and m is the mass of a specimen.

Derivation of the further relationships will be carried by the mentioned equations and numerical data.

3 Results and Discussion

As local material redistributions, connected with quite small (meso-) spatial areas (i.e. free volume), can lead to generation and growth of the STZ or shear bands during the experiment, sensitive of the proposed model to these processes must be estimated. One of the most useful parameters for describing the atomic rearrangements in MG is the thermal expansion coefficient (CTE) [14]. In the main defining points (i.e. glass-transition T_g and crystallization T_x temperatures), this parameter changes its magnitude on several whole fractions of a number in frames of order, and that identifies variation of free volume [15]. Moreover, its estimation will give the value, after that, elongation can be distinguished by mechanical loading and heating (i.e. bottom limit for analysis of nonlinear thermo-mechanical impact).

To determine the CTE, we use the standard equation for one-dimensional case (CLTE) [1]:

$$\alpha L = \frac{1}{l_0}\left(\frac{\partial l(T,\ F)}{\partial T}\right)_{F=F_{\text{load}}} \tag{4}$$

Note that it can be transformed to the derivative of Eq. (1) up to $1/l_0$ with using of (3) because Eqs. (1) and (2) can be expressed relative to each other as the two parameter functions. That is, from the system (1) and (2), the $l(T;F)$ function, whose differential equals only to $dT\left(\frac{\partial l(T,F)}{\partial T}\right)_{F=F_{load}}$ at the constant F and $dF\left(\frac{\partial l(T,F)}{\partial T}\right)_T = 0 \Leftrightarrow dF = 0$, can be expressed. The $dF = 0$ condition takes place near room temperature, where thermo-mechanical contribution to deformation will not be sufficient (only heating can be accounted). Upper bound of the quasi-constant F can be chosen as α_L ~$6 \cdot 10^{-6}$ 1/K [16] experimental value. Then, from the differential $dl(T, F_{load} = \text{Const}) = dl = dT\left(\frac{\partial l(T,F)}{\partial T}\right)_{F=F_{load}}$, we can find the necessary equation $\frac{dl}{dT}\left(\frac{\partial l(T,F)}{\partial T}\right)_{F=F_{load}}$, whose dividing on l_0 gives an alternative form of Eq. (4) from Eq. (1).

In Fig. 2, the α_L curves, calculated from the experimental data of T, are depicted with account of the analytical derivation of Eq. (1).

As it follows from our presentation and other literature, initial calculated CLTE ($4 \cdot 10^{-6}$ 1/K) agrees by magnitude order with experimental parameter both for the similar composition [16] and other ones [17]. It testifies about applicability our model for analysis of DMA on MGs. In opposite to dilatometry, $\alpha_L(T)$ curves in Fig. 2 are convex down that is caused by another loading regime. Change of C and B can be considered as function of inner structure because variation of analytical CLTE is determined only by these parameters (at the constant heating rate). Herewith, despite the numerical accordance between points in Fig. 2 and literature data (above F-bound), equation $dF = 0$ can be accounted only to F because oscillating load clearly accelerates deformation at temperature growth [11]. That is above F-line, Fig. 2 describes only the value, which is equivalent to relative deformation rate up to a constant V_T. As calculated CTE agrees with local structural properties, the further model estimation of flow and fracture is interesting.

As it was established earlier [12], deformation of amorphous alloy deviates from Newtonian behaviour, but it occurs as pseudoplastic flow [18] near fracture point.

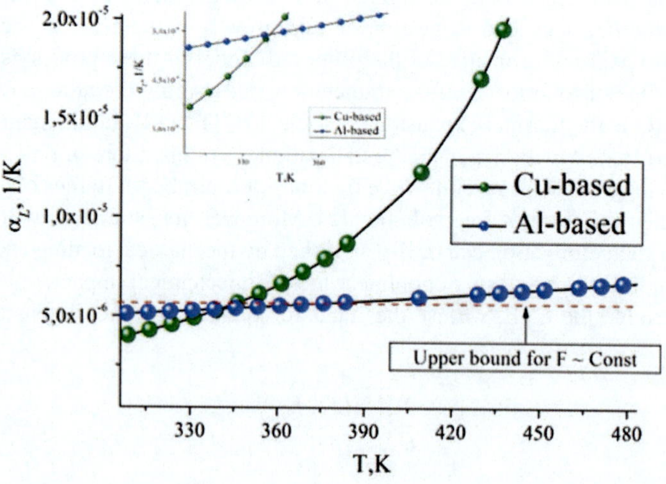

Fig. 2 Model curves of CLTE for Al-based and Cu-based ribbon amorphous alloys. The upper bound of the thermal coefficient, after that dilatometric experiment (without loading) differs from DMA, is mentioned by the arrow. Initial dynamics is presented in the insertion (upper left corner)

Therefore, the deformation can be considered in frames of hydrodynamics. The Reynolds number [19] is a criterion of nonlinear (turbulent) flow. In this case, its calculation will permit estimation of internal force dynamics and possibility of complex flow of material near fracture point. According to standard determination and Eqs. (1)–(3), Reynolds number can be presented in the form:

$$\mathrm{Re} = \frac{pvb}{\eta} = \frac{mvb}{v\eta} = \frac{mv(t(T))}{al(t(T))\eta(T)} \tag{5}$$

where a and b are thickness and width of a specimen, consequently ($a \approx$ Const); ρ is density of a ribbon, v or $v(T)$ is deformation rate (derivation of Eq. (1) with the further change of t by T), η is Newtonian dynamic viscosity, whose calculation is useful up to ~ 510 K because of the further non-Newtonian flow, V is volume of the specimen.

By using of Eqs. (1)–(3) and Eq. (5), we obtain a graphical view of the Re(T) function for Al-based and Cu-based metallic glasses (see Fig. 3).

As investigated process differs (by chemical composition and geometry) from the cases of liquids and gases, which flow occurs in the tubes or around the obstacles, absolute Re number can be another (not ~2300). But its relative form and functional growth can be used for indication of nonlinear flow at DMA. As it follows from the whole graphical view, the rate of Re(T) notably increases above 450 K that can testify about more turbulent flow of Cu-based specimen. It is well proved with SEM in Fig. 4.

There are distortions of the laminar layers near the fracture area of the ribbon. In the bottom right corner (of Fig. 4), the tendency for convergence of the curvilinear layers into the point is observed (and mentioned by the blue arrow). Note that critical

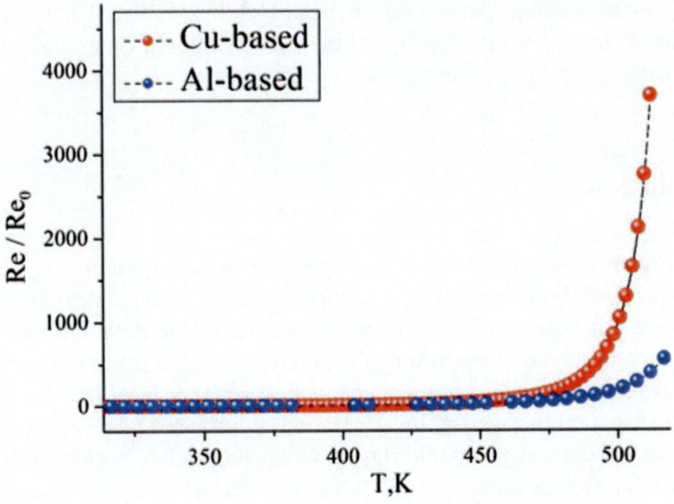

Fig. 3 Temperature relationships of the Reynolds number for investigated alloys

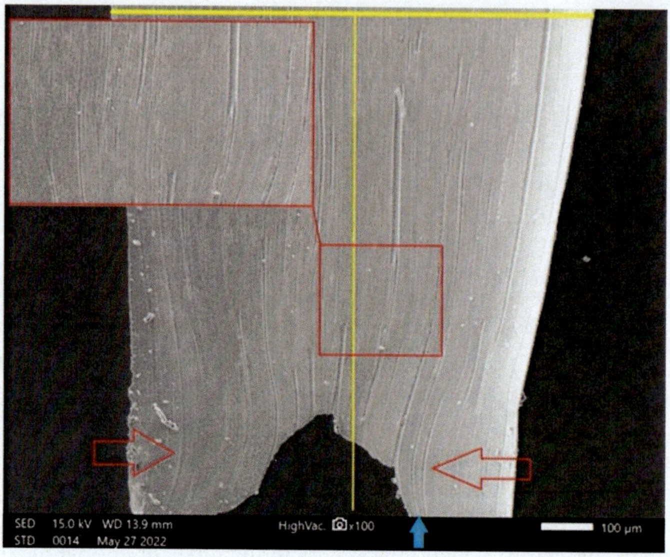

Fig. 4 SEM picture of the deformed Cu-based specimen. The most curved areas are mentioned by the red arrows or frame with zoom (upper left corner). By the yellow line, the longitudinal axis is set relatively to the wide (a least deformed) side of the specimen

Re number of the turbulent regime in metallic glass must be specified empirically in future.

As pseudoplastic viscosity of metallic glasses can be related with much redistribution of the shear bands [12], near the fracture point, and these defects can form the turbulent layers with notable boundaries (like in Fig. 4). However, in opposite to the main (pure amorphous) state, crystal nuclei (see Fig. 5) like the crystallites on the shear bands in mono- or poly- crystalline alloys [20] will occur along the layer boundaries and, probably, in other areas.

4 Conclusion

Finally, estimation of some deformation parameters was carried on $Al_{85}Y_8Ni_5Co_2$ and $Cu_{54}Pd_{28}P_{18}$ MGs in frames of non-isothermal model. Magnitudes of CLTE were calculated and compared with experiment. Dynamics of the Reynolds number was calculated for amorphous alloys as function of temperature. It is shown that turbulent elements take place for alloy with higher functional rate of Re(T), and it is proved by SEM. A structural mechanism for generation of turbulent flow is proposed with account of typical defects of metallic glasses. Obtained results can be used for further analysis of the different properties of MGs and materials with the same behaviour under identical external conditions.

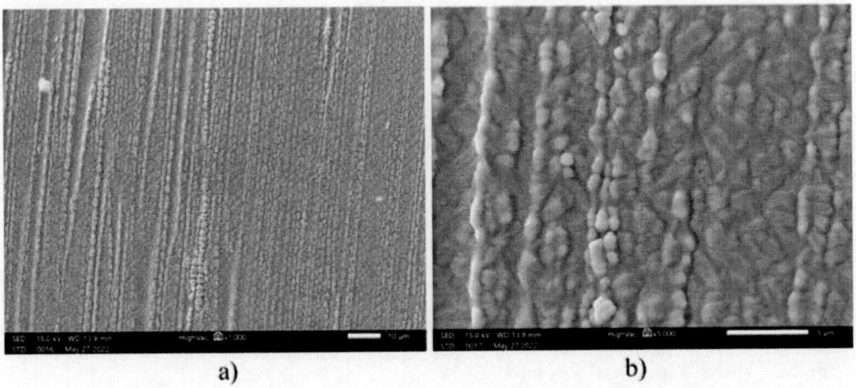

Fig. 5 Structure of the layer flow in $Cu_{54}Pd_{28}P_{18}$ MG: **a** main view; **b** enlarged area with several layers

Acknowledgements This work was supported by the Russian Science Foundation (grant No. 22-22-00226) and by the Ministry of Science and Higher Education of the Russian Federation in the frame work of agreement № 075-15-2021-709 (unique project identifier RF----2296.61321X0037). The results were partially obtained using the equipment of the Center for Collective Use of Scientific Equipment of Derzhavin Tambov State University.

References

1. Whitman AM (2019) Thermodynamics: basic principles and engineering applications. Springer
2. Barber DJ (1990) Regimes of plastic deformation—processes and microstructures: an overview. In: Deformation Processes in minerals, ceramics and rocks. The mineralogical society series, vol 1. Springer
3. Fedorov VA, Berezner AD, Beskrovnyi AI, Neov D (2018) Determining the form of a hydrodynamic flow upon creep of an amorphous cobalt-based metal alloy in a variable temperature field. Tech Phys Lett 44:678
4. Betten J (2004) Creep mechanics, In: 2nd Edition. Springer
5. Menard KP (1999) Dynamic mechanical analysis: a practical introduction. CRC Press, p 205
6. Adams BL, Field DP (1991) A statistical theory of creep in polycrystalline materials. Acta Metall Mater 39:2405
7. Hosseini ZS, Dadfarnia M, Somerday BP, Sofronis P, Ritchie RO (2018) On the theoretical modelling of Fatigue crack growth. J Mech Phys Solids 121:341
8. Tomita Y, Adachi T, Tanaka S (1998) Computational prediction in glassy polymers. Achives Comput Methods Eng 5:167
9. Abrosimova G, Aronin A, Fokin D, Orlova N, Postnova E (2019) The decrease of Young's modulus in shear bands of amorphous $Al_{87}Ni_8La_5$ alloy after deformation. Mater Lett 252:114
10. Yu HB, Wang WH, Bai HY, Wu Y, Chen MW (2010) Relating activation of shear transformation zones to β relaxations in metallic glasses. Phys Rev B 81:220201(R)
11. Berezner AD, Fedorov VA, Zadorozhnyy MY, Golovin IS, Louzguine-Luzgin DV (2022) Deformation of Cu-Pd-P metallic glass under cyclic mechanical load on continous heating. Theor Appl Fract Mech 118:103262

12. Berezner AD, Fedorov VA, Zadorozhnyy MY (2022) Relaxation behaviour of an Al-Y-Ni-Co metallic glass in as-prepared and cold rolled state. J Alloys Compd 923:166313
13. Berezner AD, Fedorov VA, Zadorozhnyy MY, Golovin IS, Louzguine-Luzgin DV (2021) Deformation of $Al_{85}Y_8Ni_5Co_2$ metallic glasses under cyclic mechanical load and uniform heating. Metals 11:908
14. Porscha B, Neuhäuser H (1994) Combined measurements of length and modulus change and calculation of effective pair potentials for the amorphous alloy $Cu_{64}Ti_{36}$. Phys Status Solidi B 186:119
15. Hu Q, Zeng H, Fu MW (2012) Characteristic free volume change of bulk metallic glasses. J Appl Phys 111:083523
16. Li GH, Wanga WM, Biana XF, Zhanga JT, Lia R, Qina JY (2009) Correlation between thermal expansion coefficient and glass formability in amorphous alloys. Mater Chem Phys 116:72
17. Su C, Niu Y, Li H, Huang G, Wang X (2011) Thermal expansion behavior of amorphous Fe-Si-B alloys. J Iron Steel Res Int 18(6):74
18. Wilkinson W (1960) Non-Newtonian fluids. Pergamon, London
19. Reynolds O (1883) An experimental investigation of the circumstances which determine whether the motion of water shall be direct or sinuous, and of the law of resistance in parallel channels. Phil. Trans. R. Soc. 174935
20. Duckham A, Engler O, Knutsen RD (2002) Moderation of the recrystallization texture by nucleation at copper-type shear bands in Al–1Mg. Acta Mater 50:2881

Finite Element Failure Analysis of Single-Lap Bolted Connection Under Impact Load

Mingpo Zheng, Zhifeng Liu, and Magd Abdel Wahab

Abstract Bolted connections are widely used in various mechanical structures due to their structural and strength advantages. However, the bolted connection will fracture and lead to failure under the action of excessive load, which affects the service reliability of the structure. In this work, the mechanical performance of single-lap bolted connection under impact load was elaborated by finite element analysis. In order to obtain a more reliable connection, different parameters were considered, i.e. tightening torque, load amplitude, friction coefficients for the contact surfaces, etc. In order to investigate the effect of various friction coefficients on the structural slip under impact load, orthogonal tests were designed and carried out. The results showed that the influence ranking of each friction coefficient on the structural slip was as follows: contact friction coefficient, thread friction coefficient, head friction coefficient, and bearing surface friction coefficient. The performance under impact load was positively related to the magnitude of the tightening torque. Within a certain impact load amplitude threshold, the structural performance was relatively stable. The reference basis for the reliability design of bolted connections based on the service and structural performance can be obtained from the finite element analysis results.

Keywords Bolted connection · Slip distance · Impact load · Reliability design

M. Zheng (✉)
Beijing Key Laboratory of Advanced Manufacturing Technology, Beijing University of Technology, 100 Pingleyuan, Chaoyang District, Beijing 100124, China
e-mail: zmp@emails.bjut.edu.cn

Z. Liu
Key Laboratory of CNC Equipment Reliability, Ministry of Education, Jilin University, No. 2699 Qianjin Street, Changchun City 130012, China

M. Zheng · M. Abdel Wahab
Laboratory Soete, Faculty of Engineering and Architecture, Ghent University, Technolgiepark Zwijnaarde 46, B-9052 Zwijnaarde, Belgium

M. Abdel Wahab (ed.), *Proceedings of the 10th International Conference on Fracture Fatigue and Wear*, Lecture Notes in Mechanical Engineering,
https://doi.org/10.1007/978-981-19-7808-1_2

1 Introduction

Bolted joint is one of the most widely used connections in various mechanical structures. However, under external loads, threaded fasteners will be self-loosening and lead to performance degradation of the connection structure during its entire service life [1]. According to the mechanical performance requirements, bolted joints were divided into different categories [2, 3]. For some applications, slippage between bolted joints is prohibited, while, in other applications, it is considered to be failure when fracture occurs. Therefore, the design and use of reasonable and reliable bolted joints should be carried out according to the specific structural form and service performance requirements.

Refs. [4–6] showed that the mechanical properties of bolted joints under transverse loads exhibited staged characteristics. Furthermore, various mechanical properties of bolted joints were exhibited under different load forms and influencing factors. Abd-Elhady et al. [7] studied the effect of structural dimensions and tightening torque on the performance of bolted joints. Jalali et al. [8] found that the surface roughness of material also affected the tangential stiffness of the structure. Izumi et al. [9] found that the transverse load threshold when large slip occurs was positively related to the clamping force and friction coefficient. Subsequently, their finite element analysis results showed that when the external load did not exceed a certain limit of the slip threshold, the mechanical properties of the structure could remain relatively stable [10].

Various service conditions of bolted joints lead to different kinds of loads and failure modes. The form of load also inevitably affects the mechanical properties of the connection. Under impact loads, larger amplitudes may sometimes be reached in a short period of time, making the bolted joint more prone to failure. Herein, finite element analysis method is adopted to investigate the mechanical properties of bolted joints under impact load. Orthogonal tests are designed and carried out to investigate the effect of various friction coefficients of bolted joint on structural slip under transverse impact load. In addition, the displacement under different tightening torques and load amplitudes is also investigated. The reliability design of bolted joint can be carried out through the finite element analysis results under various influencing factors.

2 Finite Element Analysis Setup

2.1 Load and Boundary Conditions

In the finite element analysis, two plates made of the same dimensions were joined together by M16 bolts and nuts. The length and width of the connecting plate were 80 mm and 60 mm, respectively. The bolt hole diameter was 17 mm. There are a total of three analysis steps that were set during the finite element analysis. In

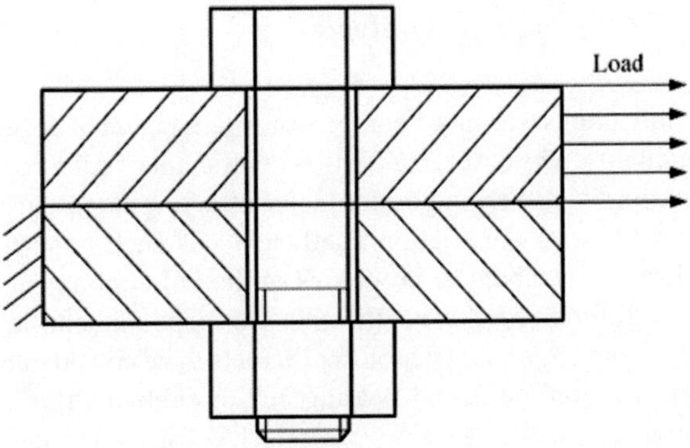

Fig. 1 Load and boundary conditions during loading step

the first analysis step, the tightening torque was applied to obtain the bolt preload. Afterwards, in the second analysis step, the loads and boundary conditions of the bolted connection were removed to simulate the service situation after tightening. In the third analysis step, a transient impact load was applied to the right side of the upper plate. Meanwhile, the left side of the lower plate was fully constrained. The schematic diagram of the load and boundary conditions of the single-lap bolted connection structure is shown in Fig. 1. The displacement at the loading point in the third analysis step was monitored for analysis and evaluation of the mechanical properties of the structure.

As shown in Fig. 1, the bolt connection consists of four parts. The hexahedral element global mesh size of the two connecting plates was 2 mm. The local mesh size was refined at the bolt hole locations. The hexahedral mesh element of the threaded fastener had a global size of 1 mm. Among them, there are only tetrahedral mesh elements at the thread, with which the local size was 0.5 mm. There were a total of 25,325 hexahedral elements and 119,586 tetrahedral elements for bolted joint. Six contact pairs are set for the four parts, namely: the contact between the bolt head and the upper plate (head friction coefficient, μ_h), the contact between the two connecting plates (contact friction coefficient, μ_m), the contact between the lower plate and the nut (bearing friction coefficient, μ_b), the contact surface between threads (thread friction coefficient, μ_t), and the contact between the bolt hole in the two plates and the outer surface of the bolt. In the first analysis step (tightening step), different tightening torques were applied to analyze the mechanical properties under different clamping forces, ranging from 150 to 210 Nm with 15 Nm intervals. In addition, the magnitude of the instantaneous impact load was also considered. The load amplitudes applied in the third analysis step (loading step) ranged from 5 to 25 kN at 5 kN intervals.

2.2 Design of Orthogonal Analysis

In engineering applications, various manufacturing parameters of bolted joints lead to different surface quality. Moreover, the friction coefficient of each contact surface of the bolted connection is also an important factor affecting its resistance to external loads. Orthogonal tests on the friction coefficients of each contact surface were designed to explore thc mechanical properties of the bolted joints under transverse impact loads under different friction coefficients. Corresponding orthogonal table are shown in Table 1. There were four friction coefficients to be considered: head friction coefficient, contact friction coefficient, bearing friction coefficient, and thread friction coefficient. Among them, four levels were selected for each friction coefficient, ranging from 0.10, 014, 0.18, and 0.22. The orthogonal test category was $L16$, with four factors and four levels. Therefore, a total of 16 tests of finite element analysis were carried out. Since under large transverse loads, the plate where the load was applied may come into contact with the bolt, the contact of the bolts with the bolt holes of the two plates was set with a friction coefficient of 0.15.

Table 1 Orthogonal experimental design table

S. No.	Friction coefficients of the structure			
	μ_h	μ_m	μ_b	μ_t
1	0.10	0.10	0.10	0.10
2	0.10	0.14	0.14	0.14
3	0.10	0.18	0.18	0.18
4	0.10	0.22	0.22	0.22
5	0.14	0.10	0.14	0.18
6	0.14	0.14	0.10	0.22
7	0.14	0.18	0.22	0.10
8	0.14	0.22	0.18	0.14
9	0.18	0.10	0.18	0.22
10	0.18	0.14	0.22	0.18
11	0.18	0.18	0.10	0.14
12	0.18	0.22	0.14	0.10
13	0.22	0.10	0.22	0.14
14	0.22	0.14	0.18	0.10
15	0.22	0.18	0.14	0.22
16	0.22	0.22	0.10	0.18

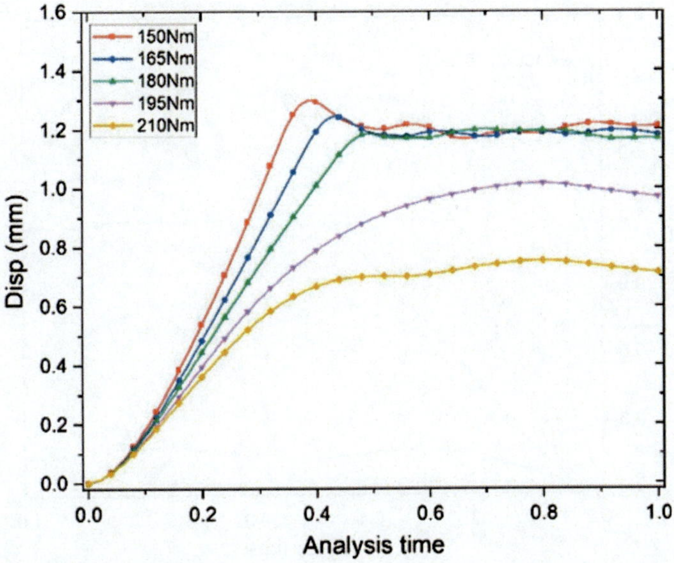

Fig. 2 Slip distance of bolted connection under impact load with different tightening torque

3 Finite Element Analysis Results

3.1 *Tightening Torque*

The finite element analysis results obtained under different tightening torques are shown in Fig. 2. The magnitude of the impact load applied was 15 kN. In the loading step, the transverse impact load reached the pre-set value at the moment of initial application, and the corresponding acceleration was generated. Therefore, the maximum slip distance could only be reached within a certain period of time after the load was applied. It can be seen from the figure that the displacement amplitude under impact load was negatively correlated with the tightening torque. At lower tightening torques, a certain drop occurred after the maximum slip distance. However, under the larger tightening torque, the displacement after reaching the maximum displacement amplitude remained relatively stable. It showed that the ability to resist transverse impact load could be improved by means of increased tightening torque.

3.2 *Load Amplitude*

To investigate the effect of transverse impact load amplitude on the mechanical properties of bolted connections, different load amplitudes were used in this work. The

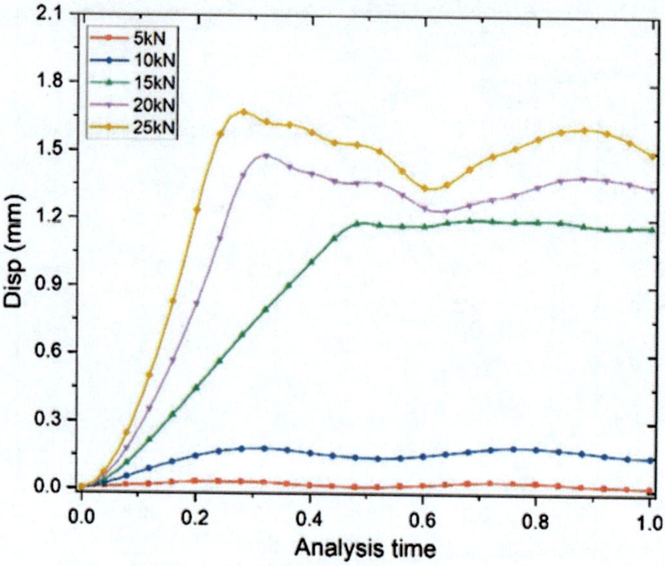

Fig. 3 Slip distance of bolted connection under impact load with different load amplitude

same tightening torque of 180 Nm was selected in the finite element analysis. Similarly, after reaching the maximum slip distance under heavier loads, it subsequently decreased. When the load did not exceed 15 kN, it remains relatively stable after reaching the maximum slip distance. The slip distance did not exceed 0.3 mm under the action of the load amplitude of 10 kN. When the load was reduced to 5 kN, the corresponding slip distance was almost negligible. It should be noted that when the load amplitude was increased from 10 to 15 kN, the generated slip distance rapidly increased to a larger amplitude. It showed that there was a certain threshold value below which the external impact load can remain stable (Fig. 3).

3.3 Orthogonal Test Results

The finite element analysis was carried out according to the orthogonal series in Table 1. The results obtained were shown in Fig. 4. In order to avoid the influence of the clamping force amplitude on the analysis results, the clamping force was controlled. The average clamping force was 55.44 kN (the error does not exceed $\pm 1.1\%$). It could be seen from the figure that the largest slip distance was in the first group of tests, in which the friction coefficient of each contact surface was at a low level (0.10). This was consistent with the expected result that a smaller coefficient of friction is detrimental to transverse load resistance. The first four groups of tests with larger slip

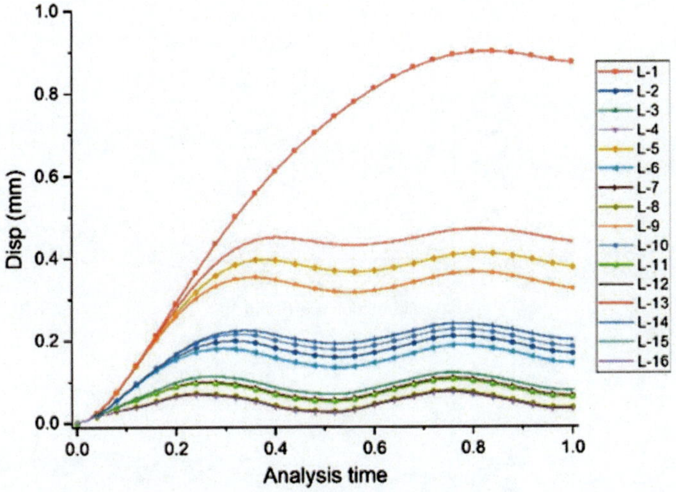

Fig. 4 Orthogonal test results under impact load

distance in the figure were when the friction coefficient of the contact surface was smaller. It showed that the friction coefficient of the contact surface had the greatest influence on the slip distance of the structure.

According to the obtained orthogonal test results, the analysis results of each influencing factor and its level are shown in Fig. 5. Similar to the results in Fig. 4, the slip distance is negatively correlated with the contact friction coefficient. The results showed that when the friction coefficient of the bearing surface was increased, the slip distance can be significantly reduced. While for the friction coefficients of other contact surfaces, when the friction coefficient was at a low level (0.10), the slip distance is larger. However, when the coefficient of friction is increased, the slip distance obtained is somewhat reduced. As it increases further, the slip distance remains relatively stable. It showed that when the friction coefficient was lower than a certain limit, the slip distance would be greatly improved. The results showed that the excessively small friction coefficient should be avoided in the design and use to improve the mechanical properties of the bolted connection.

Under the same load amplitude, the influence ranking of various factors on the slip distance is shown in Table 2. It can be seen from the table that the contact friction coefficient had the greatest influence on the slip distance. The second is the thread friction coefficient. The head friction coefficient and the bearing surface friction coefficient have less influence and are almost identical. It is because the relative rotation of the bolt will be driven when the plate on which the load is applied slips greatly. As the weak link of the structure, the threaded part will be more prone to slip.

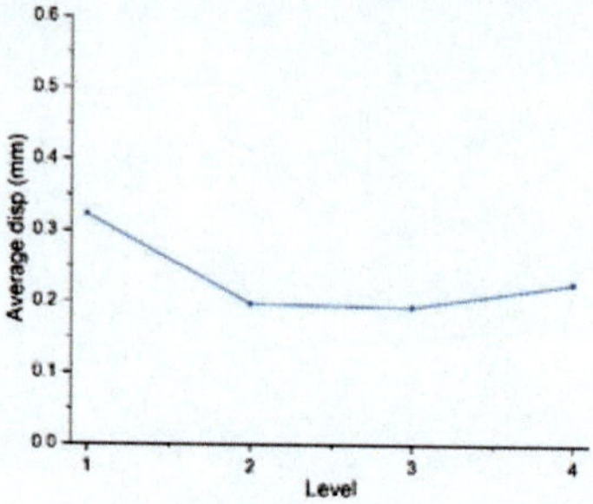

a) head friction coefficient

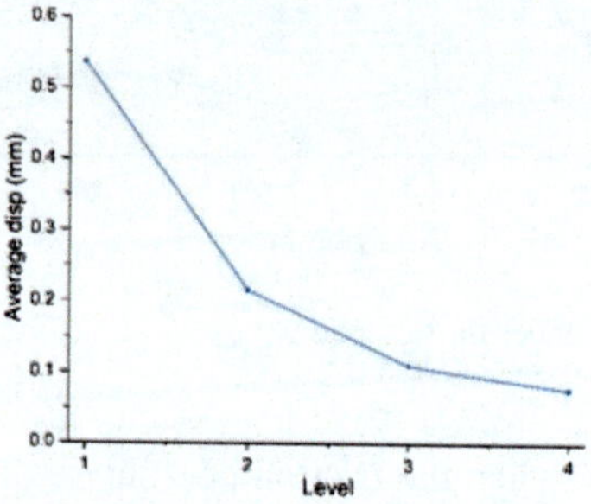

b) contact friction coefficient

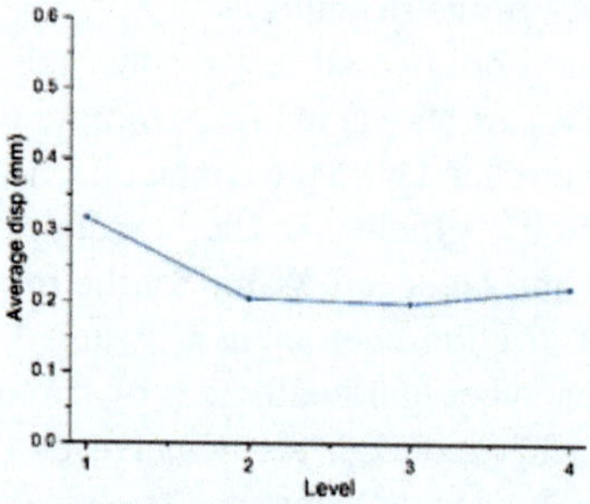

c) bearing friction coefficient

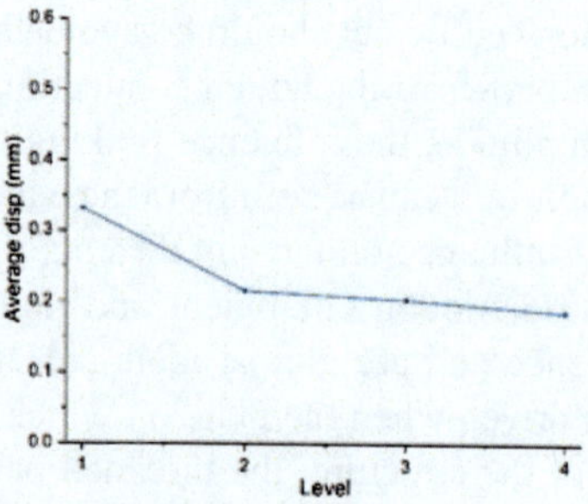

d) thread friction coefficient

Fig. 5 Orthogonal test analysis results, **a** head friction coefficient, **b** contact friction coefficient, **c** bearing friction coefficient, **d** thread friction coefficient

Table 2 Orthogonal test results for friction coefficients

Conditions	μ_h	μ_m	μ_b	μ_t
Delta	0.1296	0.4607	0.1193	0.1444
Rank	3	1	4	2

4 Conclusions

Through the designed finite element analysis orthogonal test, the influence of the slip distance of each contact surface of the bolted joint was obtained, followed by, contact friction coefficient, thread friction coefficient, head friction coefficient, and bearing friction coefficient. Increasing the tightening torque improved the resistance to transverse impact loads. Under certain service conditions, when the load did not exceed a certain threshold, the slip distance was relatively stable. The finite element analysis results carried out herein can provide reference values for the reliability design of single-lap bolted joints.

Acknowledgements The authors gratefully acknowledge the financial support provided by the National Natural Science Foundation of China (No. 51975019).

References

1. Bickford JH (2007) Introduction to the design and behavior of bolted joints, 4th edn. Marcel Dekker, New York
2. BS EN 1993-1-8 (2005) Eurocode 3: design of steel structures—Part 1–8: design of joints
3. BS EN 1990 (2002) Eurocode—basis of structural design
4. Khashaba UA, Sallam HEM, Al-Shorbagy AE, Seif MA (2006) Effect of washer size and tightening torque on the performance of bolted joints in composite structures. Compos Struct 73(3):310–317
5. Al-Nassar YN, Khurshid H, Arif AFM (2012) The effect of clearance and pre-tension on the performance of a bolted-joint using 3D FEA. Arab J Sci Eng 37(3):749–763
6. Zhifeng L, Mingpo Z, Jinfei G, Hongyan C, Xing Y, Ying L (2021) Experimental study on performance characterization of bolted joint under transverse loading. Measurement 2021(182):109608
7. Abd-Elhady A, Abu-Sinna A, Atta M, Sallam HEDM (2020) Identification of damage stages in bolted metallic joints for different joint geometries and tightening torques using statistical analysis. Adv Struct Eng 23(5):911–923
8. Jalali H, Khodaparast HH, Friswell MI (2019) The effect of preload and surface roughness quality on linear joint model parameters. J Sound Vib 447:186–204
9. Izumi S, Yokoyama T, Iwasaki A, Sakai S (2005) Three-dimensional finite element analysis of tightening and loosening mechanism of threaded fastener. Eng Fail Anal 12(4):604–615
10. Izumi S, Kimura M, Sakai S (2007) Small loosening of bolt-nut fastener due to micro bearing-surface slip: A finite element method study. J. Solid Mech. Mater. Eng. 1(11):1374–1384

Plasticity Analysis in Aluminum Alloy Plates Repaired with Bonded Composite Patch Under Overload

Check for updates

Faraz Ahmed, S. M. A. K. Mohammed, Faycal Benyahia, Bel Abbes Bachir Bouiadjra, and Abdulmohsen Albedah

Abstract In this study, we analyzed the effect of plasticity on the fatigue crack growth of repaired aluminum plate bonded with composite patch subjected to single overload. Experimental and numerical approaches were used to perform the analysis. In the experimental part, fatigue tests were carried out on a v-notched aluminum plate repaired with a composite patch and subjected to overload. In the numerical part, the extent of the plastic zone at the crack tip of the repaired and unrepaired cracks were computed. The obtained results showed that the effect of overload compared to that of the composite patch led to a remarkable improvement of the fatigue life.

Keywords Bonded composite repair · Fatigue life · Overload · Crack growth · Plastic zone

---

F. Ahmed · F. Benyahia · A. Albedah (✉)
Mechanical Engineering Department, College of Engineering, King Saud University, Riyadh, Saudi Arabia
e-mail: albedah@ksu.edu.sa

F. Ahmed
e-mail: faraz385@gmail.com

F. Benyahia
e-mail: fbenyahia@ksu.edu.sa

S. M. A. K. Mohammed (✉)
Department of Mechanical and Industrial Engineering, Ryerson University, Toronto Ontario M5B 2K3, Canada
e-mail: sohail@ryerson.ca

B. A. B. Bouiadjra
Department of Mechanical Engineering, LMPM, University of Sidi Bel Abbes, BP 89, Cité Ben M'hidi, 22000 Sidi Bel Abbes, Algeria
e-mail: bachirbou@yahoo.fr

M. Abdel Wahab (ed.), *Proceedings of the 10th International Conference on Fracture Fatigue and Wear*, Lecture Notes in Mechanical Engineering,
https://doi.org/10.1007/978-981-19-7808-1_3

1 Introduction

The application of overload during a cyclic loading causes a retardation in the propagation of the crack. The retardation could last for very large number of cycles depending on the material type, overload magnitude, loading condition, etc. [1–4]. Such a retardation is attributed to the formation of plastic zone and blunting of the crack tip. The energy requires for the crack to propagate following the crack blunting increases dramatically. However, after a threshold overload amplitude, the crack retardation will not be effective as the damage could be severe, leading to catastrophic failure [5–7]. Under plane stress conditions, the high plasticity around the crack front leads to a blunting of the crack front for aerospace grade aluminum alloys such as Al2024, Al7075, etc. [8, 9]. As the blunting of the crack happens, the effective stress intensity factor (SIF) decreases.

Bonded composite repair is a widely accepted technology for the repair of cracked metallic structures under fatigue loading [10–12]. A high-performance composite patch with much higher stiffness and strength than the substrate is bonded on to a cracked plate causing a load transfer from the weaker substrate to the stronger patch through an adhesive layer. By doing so, the crack is retarded or stopped permanently (in some cases) and eventually reduces stress intensity around the crack front leading to a global improvement in the fatigue life [13, 14]. In this present study, the combined effect of plasticity induced by overload and the bonded composite repair patch on an aluminum alloy plate will be investigated experimentally and numerically. We have calculated the extent of the plastic zone around a crack repaired with a composite patch under fatigue overload using the finite element method.

2 Experimental Setup

In the current study, Al2024-T3 specimens were used as substrate. Pre-impregnated epoxy carbon was supplied in roll form by Rock West composites. The unidirectional fiber carbon composite contains approximately 65% fiber by volume. The laminated composite was fabricated by stacking 8 unidirectional layers in the form of composite patch. This choice was dictated by the fact that the plates are subjected to uniaxial fatigue tests, the direction of the fibers must be parallel to the direction of loading to have maximum efficiency of the repair.

The V-notch fatigue specimens were machined on a waterjet cutter. The base and height of the notch is equal to 6 mm with an angle of 60°. Fatigue tests were conducted on the specimens prior to repair to obtain a 3 mm pre-crack. The unidirectional carbon/epoxy patch was applied on the cracked area using Araldite 2015 epoxy (see Fig. 1).

Fatigue tests were performed on a 100 kN Instron fatigue machine (model 8801) under load-controlled mode with a stress ratio of 0.1. On both repaired and unrepaired aluminum plates, the maximum stress was maintained at 70 MPa. As the crack

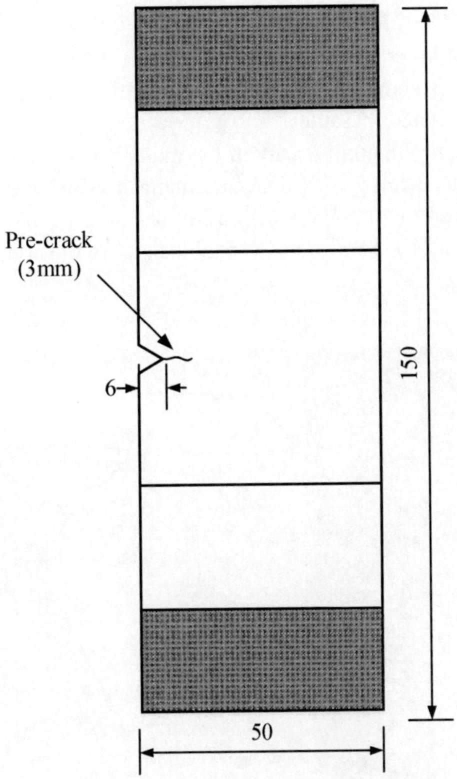

Fig. 1 Geometry of the cracked Al plated repaired with carbon/epoxy composite patch

reached 3 mm, a single tensile overload of magnitude 140 MPa with an overload ratio of 2 was applied. The constant amplitude loading was then continued until the specimen failed completely. A digital camera was used to monitor the fatigue crack growth rate (FCG).

3 Numerical Model

The Ansys computational code was used to achieve the numerical part of this study. The finite element model consists of three sub-layers that respectively model the cracked plate, the adhesive and the composite patch. The plate, the adhesive and the composite are meshed separately with 20-node brick elements with the same mesh on the contact surfaces. In order to have an accurate calculation at the crack front, we refined the mesh around the front. The total number of elements employed in this research is 17028: 12,216 for the aluminum plate, 3208 for the composite patch and 1604 for the adhesive layer. The finite element model (FEM) for the overall structure and the crack region is shown in Fig. 2. The mechanical properties of the materials used in the study are listed in Table 1.

To take into account the nonlinearity of the materials in the FEM, the Von-Mises criterion was used associated with the incremental plasticity theory. By using the Newton - Raphson iterative method with a limited number of steps of 100 and an increment size between 10, 5 and 1, we were able to solve the nonlinear finite element equations.

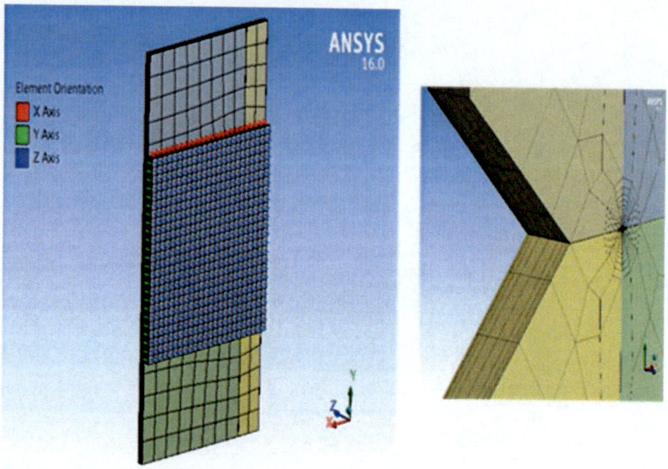

Fig. 2 Typical mesh model of the assembly and near the crack tip [13]

Table 1 Elastic properties of materials used in this study

Properties	Material		
	Al 2024-T3	Carbon/epoxy	Adhesive (Araldite 2015)
Longitudinal young's modulus (GPa)	72.4	130	2.52
Transversal young's modulus (GPa)	72.4	9	2.52
Longitudinal poisson ratio	0.33	0.33	0.36
Transversal poisson ratio	0.33	0.03	0.36

4 Results and Discussion

4.1 Experimental Results

Figure 3 shows the FCG curves ($da/dN = f(a)$) for three cases, i.e., constant amplitude loading (CAL), CAL followed by an OL at a crack length of 3 mm (CAL + OL + CAL) and CAL followed by an OL at a 3 mm repaired with a composite patch (CAL + OL + P + CAL). The CAL curve is pretty much linear as the crack length progressed. In contrast, an abrupt dip in the crack growth rate of two other cases (CAL + OL + CAL and CAL + OL + P + CAL) was observed at a crack length of 3 mm where the overload was applied. Such a dip exemplifies the crack retardation which significantly improves the fatigue life of the structures. The decrease in crack growth of OL + patch is higher than the OL alone which demonstrates the combined effect of OL and patch repair. This proves that by bonding the patch prior to the overload, both retardation effects are added, and the fatigue life is greatly improved. The application of the patch overload has a significant impact on the fatigue life extension [15, 16]. As a result, after applying overload, it is recommended to repair the cracked specimen using a composite patch, which can result in a significant increase in fatigue life when compared to the unrepaired specimen.

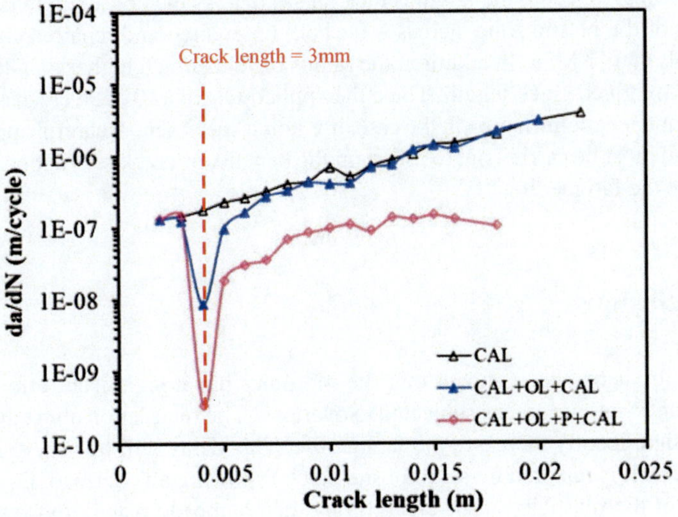

Fig. 3 Crack growth curves for different cases investigated

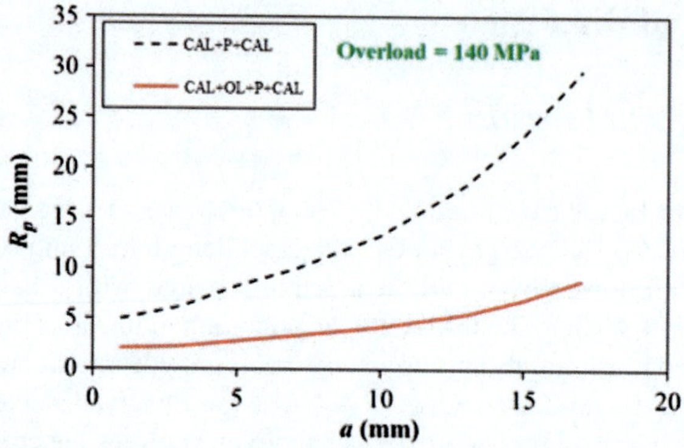

Fig. 4 Variation of the plastic zone radius for repaired and unrepaired plates evaluated numerically

4.2 Numerical Results

Finite element analysis was performed using Ansys to determine the degree of plastic zone formed at the crack tip due to a single tensile overload during a fatigue loading. In order to demonstrate the contrast, the radius of plastic zone was calculated for unrepaired condition as well. Figure 4 shows the variation of the radius of plastic zone (R_p) with respect to the fatigue crack length (a). As the crack length increases, the radius of the plastic zone increases for both unrepaired and repaired conditions with an OL of 140 MPa. In addition, the radius of R_p is much higher than the crack length at any given crack length. Thus, the applied OL of 140 MPa creates a much higher permanent deformation in the case of v-notch specimen. A careful application of overload must be carried out to ascertain the benefits of crack retardation in order to improve the fatigue life.

5 Conclusion

In this study, it has been shown that the plasticity has a significant effect in the propagation of fatigue cracks subjected to overload. The retardation effect due to the overload significantly increases the fatigue life. This delay will be very significant when the crack is repaired by a composite patch following an overload. Despite the reduction of the plasticity at the crack tip by the composite patch, it also causes a retardation in the crack propagation. The two combined retardation effects (patch + overload) results in a virtually infinite fatigue life of the metallic structure repaired by a composite patch.

Acknowledgements This Project was funded by the National Plan for Science, Technology and Innovation (MAARIFAH), King Abdulaziz City for Science and Technology, Kingdom of Saudi Arabia, Award Number (13-ADV2167-02).

References

1. Albedah A, Bouiadjra BB, Mohammed S, Benyahia F (2019) Fractographic analysis of the overload effect on fatigue crack growth in 2024–T3 and 7075–T6 Al alloys. Int J Miner Metall Mater 27:83–90
2. Schijve J (2015) The application of small overloads for fractography of small fatigue cracks initiated under constant-amplitude loading. Int J Fatigue 70:63–72
3. Celik CE, Vardar O, Kalenderoglu V (2004) Comparison of retardation behaviour of 2024–T3 and 7075–T6 Al alloys. Fatigue Fract Eng Mater Struct 27:713–722
4. Verma BB, Ashok K, Ray PK (2000) Fatigue crack growth delay following overload. Trans Indian Inst Met 53:591–595
5. Bouchkara NHM, Albedah A, Benyahia F, Mohammed SMAK, Bouiadjra BB (2021) Experimental and numerical analyses of the effects of overload on the Fatigue life of aluminum alloy panels repaired with bonded composite patch. Int. J. Aeronaut Space Sci 1–10
6. McEvily AJ, Ishikawa S, Makabe C (2002) The influence of the baseline R value on the extent of retardation after an overload. Mechanism and mechanics of fracture. In: The John Knott symposium, pp 37–42
7. Mohammed SMAK, Bouiadjra BB, Benyahia F, Albedah A (2018) Analysis of the single overload effect on fatigue crack growth in AA 2024–T3 plates repaired with composite patch. Eng Fract Mech 202:147–161
8. Li S, Zhang Y, Qi L, Kang Y (2018) Effect of single tensile overload on fatigue crack growth behavior in DP780 dual phase steel. Int J Fatigue 106:49–55
9. Fekih SM, Albedah A, Benyahia F, Belhouari M, Bouiadjra BB, Miloudi A (2012) Optimization of the sizes of bonded composite repair in aircraft structures. Mater Des 41:171–176
10. Mohammed S, Mhamdia R, Albedah A, Bouiadjra BAB, Bouiadjra BB, Benyahia F (2021) Fatigue crack growth in aluminum panels repaired with different shapes of single-sided composite patches. Int J Adhes Adhes 105:102781
11. Rose LRF, Wang CH (2002) Analytical methods for designing composite repairs. In: Baker A, Jones R, Rose LRF (eds) Advances in bonded composite repairs of metallic aircraft structure. Elsevier, Amesterdam, pp 137–175
12. Serier B, Bouiadjra BB, Belhouari M (2012) Numerical analysis of the patch shape effects on the performances of bonded composite repair in aircraft structures. Compos B Eng 43:391–397
13. Demmouche N, Albedah A, Mohammed SMAK, Bouiadjra BB, Benyahia F (2019) Interaction between adherend plasticity and adhesive damage in metal/composite joints: application to bonded composite repair of metallic structures. Weld World 63(2):211–219
14. Brighenti R, Carpinteri A, Vantadori S (2006) A genetic algorithm applied to optimisation of patch repair for cracked plates. Comput Methods Appl Mech Eng 196:466–475
15. Mohammed SMAK, Albedah A, Benyahia F, Bouiadjra BB (2018) Effect of single tensile peak overload on the performance of bonded composite repair of cracked Al 2024-T3 and Al 7075-T6 plates. Compos Struct 193: 260–267
16. Albedah A, Khan SMA, Bouiadjra BB, Benyahia F (2018) Fatigue crack propagation in aluminum plates with composite patch including plasticity effect. Proc Inst Mech Eng G J Aerosp Eng 232(11):2122–2131

Machine Learning for Predicting Pipeline Displacements Based on Soil Rigidity

Meriem Seguini, Samir Khatir, Djamel Nedjar, and Magd Abdel Wahab

Abstract This study investigates the impact of the soil rigidity on the mechanical behaviour for linear and nonlinear pipelines. The work is based on the results of a series of mechanical finite element analyses based on the VanMarcke and Artificial Neural Network (ANN). The numerical model is validated based on the literature. Different simulations have been generated to obtain data response of the pipe based on displacement. The predicted results using ANN are compared with VanMarcke to prove the effectiveness and the importance of the ANN. The analysis proves that the variation of the coefficient of subgrade reaction can induce a significant displacement of the pipe. The results prove that ANN provides a major role in the evolution of the real displacement of the pipeline and allows us to obtain more precise and interesting results based on both linear and nonlinear cases.

Keywords Finite element analysis · Interaction soil structure · Coefficient of subgrade reaction · Numerical simulations · Static analysis · ANN · VanMarcke method

1 Introduction

The pipelines are the innovative mean of energy and water transport, for that reason the study of the interaction soil-pipeline has been very attractive. In fact, a range of studies has been performed on the interaction of soil structure [1, 2]. The influence of the soil's nonlinear behaviour on different types of structures has been also studied [3, 4].

M. Seguini (✉) · D. Nedjar
Department of Civil Engineering, Faculty of Architecture and Civil Engineering, Laboratory of Mechanic of Structures and Stability of Constructions, University of Sciences and Technology of Oran Mohamed Boudiaf, Bir El Djir, Algeria
e-mail: meriem.seguini@univ-usto.dz

S. Khatir · M. Abdel Wahab
Soete Laboratory, Faculty of Engineering and Architecture, Ghent University, Technologiepark Zwijnaarde 903, B-9052 Zwijnaarde, Belgium

M. Abdel Wahab (ed.), *Proceedings of the 10th International Conference on Fracture Fatigue and Wear*, Lecture Notes in Mechanical Engineering,
https://doi.org/10.1007/978-981-19-7808-1_4

Dynamic analysis has been also performed to obtain additional observations and results on the soil-structure interaction problems [5]. For much improvement of the structure (beam, pipeline, plate…,etc.) design, a stochastic analysis has been done by different researchers [5–13] where the soil has been considered as heterogeneous. Grigoriu et al. [12] studied the behaviour of a beam on random soil.

Griffiths et al. [13] used a probabilistic method to analyse the response of the pile supported by Winkler random soil. Recently, several other numerical models of beam and pipeline resting on linear and nonlinear soil have been developed where the VanMarcke method combined with Monte Carlo simulations has been used. Seguini and Nedjar [10, 11] developed a simplified finite element model of soil-beam interaction based on stochastic analysis in order to quantify the effect of different random soil parameters on the pipe response. This study has demonstrated the significant impact of the variation of the random soil's spatial parameters on the behaviour of the structure by the development of displacement variations along the beam length. To investigate the nonlinearity effect of the soil's on the behaviour of the beam, Seguini and Nedjar [14] performed a series of tests and simulations by using the nonlinear analysis [15, 16]. In addition, recent research has been done on soil-structure interaction including the dynamic stochastic analysis of the pipeline [17–19]. In fact, Seguini and Nedjar [19] provided more accurate numerical modelling of the pipeline involving the seismic effect.

However, other techniques for prediction such as ANN-PSO have been developed and used to identify the damage crack in different types of structures (beam and pipe) [20–23] were the obtained results prove the efficiency of the ANN-PSO method on the crack identification.

The objective of this paper is in particular to expand the influence (effects) of the variation of the soil subgrade reaction on the response of the pipeline where the study is based on the results of a series of FEM obtained by using the VanMarcke method [24] which is compared with the data generated by ANN technique. The FEM was performed by using MATLAB. In fact, the new proposed fast approach allows us to optimize the results in order to obtain a more realistic pipeline's response.

2 Deterministic Approach

However, in the following numerical analyses, the pipeline behaviour is considered to be nonlinear, whereas the soil behaves according to elastic constitutive model [10]. The numerical approach is based on the von Kàrmàn method [15, 16] which has been used to represent the nonlinear behaviour of the pipeline [24]. Therefore, the governing equation of the pipeline resting on elastic soil is defined as follows [24]:

$$E_p I_p \frac{d^4 w_0}{dx^4} - \frac{3}{2} E_p A_p \left(\frac{dw_0}{dx}\right)^2 \left(\frac{d^2 w_0}{dx^2}\right)^2 + p(x) = q(x) \quad (1)$$

The soil reaction $p(x)$ is given by the following formulation:

$$p(x) = k_{soil}.\pi.R_{ext}.w_0(x) \quad (2)$$

k_{soil} is the modulus of the subgrade reaction of soil; R_{ext} $D_{\text{ext}}D_{\text{ext}}$, Ip, and Ep are, respectively, the half-perimeter, the inertia and the Young's modulus of the pipe. $w_0(x)$ is the polynomial displacement function which depends on the vector N composed of Hermite cubic interpolations functions and on the vector of the nodal displacements Δ^e (*e* represents an element of a discretized pipeline).

The differential Eq. (1) is resolved by using the total potential energy functional:

$$\Psi = \Psi_1 + \Psi_2 \quad (3)$$

Ψ_1 and Ψ_2 are, respectively, the strain energy of the pipe and the soil.

$$\Psi_1 = \frac{1}{2}\{\Delta^e\}^t \left\{ \left[\frac{1}{4} E_p^e A_p^e \left(\frac{w_2 - w_1}{l_e}\right)^2 \int_0^{le} \left(\frac{dN}{dx}\right)\left(\frac{dN}{dx}\right)^t dx \right] + \left[E_p^e I_p^e \int_0^{le} \left(\frac{d^2N}{dx^2}\right)\left(\frac{d^2N}{dx^2}\right)^t dx \right] \right\} \{\Delta^e\} \quad (4)$$

$$\Psi_1 = \frac{1}{2}\{\Delta^e\}^t [K_{pnl}^e\{\Delta^e\} + K_{pl}^e]\{\Delta^e\} \quad (5)$$

$$\Psi_2 = \frac{1}{2}\{\Delta^e\}^t \left(k_{\text{soil}} \cdot \pi \cdot R_{\text{ext}} \cdot \int_0^{\text{le}} \{N\} \cdot \{N\}^t \text{dx}\right)\{\Delta^e\} \quad (6)$$

$$\Psi_2 = \frac{1}{2}\{\Delta^e\}^t [K_{\text{sl}}^e]\{\Delta^e\} \quad (7)$$

Finally, the following incremental Eq. (8) is solved by using Newton–Raphson method [24].

$$\left[K^e\{\Delta^e\}^i\right]\{\Delta^e\}^{i+1} = \{F\} \quad (8)$$

$$\{\Delta^e\}^{i+1} = \{\Delta^e\}^i + \{\delta\Delta\} \quad (9)$$

where $\{F\}$ and $[K^e]$ are the element force vector and the element rigidity matrix, respectively, as defined in Ref [24].

3 Probabilistic Approach

3.1 VanMarck's Method

The VanMarcke approach [25, 26] is based on the local average method where the random coefficient of soil's subgrade reaction k_{soil} is characterized by a lognormal distribution. Their local average and variance are defined, respectively, as follow:

$$E[k_{soil}(D_i)] = m_k \tag{10}$$

$$Var[k_{soil}(D_i)] = \sigma_k^2 \gamma(D_i) \tag{11}$$

where the variance function γ(Di) is written as

$$\gamma(D_i) = \frac{2}{D_i} \int_0^{D_i} (1 - \frac{x}{D_i}) \rho(x) dx \tag{12}$$

with

$$\rho(\tau) = 1 - \frac{|\tau|}{L_c} \text{ for } |\tau| \leq L_c \tag{13}$$

The discretization form of the variance function γ(Di) is defined as follow:

$$\gamma(D_i) = \begin{cases} 1 - \dfrac{D_i}{3L_c} & if\ D_i \leq L_c \\ \dfrac{L_c}{D_i}\left(1 - \dfrac{L_c}{3D_i}\right) & if\ D_i \succ L_c \end{cases} \tag{14}$$

In order to compute the covariance matrix C_{ij} of the coefficients of soil's subgrade reaction, we use the local average method defined by Fenton and VanMarcke [27].

$$C_{ij} = Cov\left[k_{soil}(D_i), k_{soil}\left(D_j\right)\right] \tag{15}$$

$$C_{ij} = \frac{\sigma_k^2}{2} \left\{ \begin{array}{l} (t-1)^2 \gamma[(t-1)D] - 2t^2 \gamma[t \cdot D] \\ + \left\{(t-1)^2 \gamma[(t-1)D] - 2t^2 \gamma[t \cdot D] + (t+1)^2 \gamma[(t+1)D]\right\} \end{array} \right\} \tag{16}$$

(i) and (j) are the zone indices where t = |i–j|. The zone's length $D = Di = Dj$.

3.2 ANN

The artificial neural network (ANN) is computational model that mimics the way nerve cells work in the human brain. Many researchers were used this method for classification, identification, control system..., etc. Therefore, the goal of this approach is to improve performance and find the best results in order to determine the real response and faster compared with FEM and analytical solution. In this work, ANN method is based on three components: an input layer, hidden layer, and output layer as represented in Fig. 1.

w_{kl} is the weights of neuron connection between input node and neuron in the hidden layers.

b_l is the bias, and w_l is the weights of neuron connection between neuron in hidden and output layers. b_1 is bias associated with the single neuron in the output layer. Indices $Q = 1, 2, \ldots, m$ is the number of collected data, and $S = 1, 2, \ldots, n$ is hidden layer neurons. The total number of parameters (weight and biases) employed in the network is $n \times (m + 2) + 1$.

Once the structure of the ANN model has been built, training with input and output sets are conducted to find the appropriate weights and biases and can be improved by optimization techniques as defined in Ref [28]. Therefore, the objective is to minimize the difference between the results obtained from VanMarcke method and ANN approach by using the root-mean-square error (RMSE) function which is defined as follows

$$\text{RMSE} = \sqrt{\frac{\sum_{s=1}^{n}(O_S - l_S)^2}{\text{nd}}} \tag{17}$$

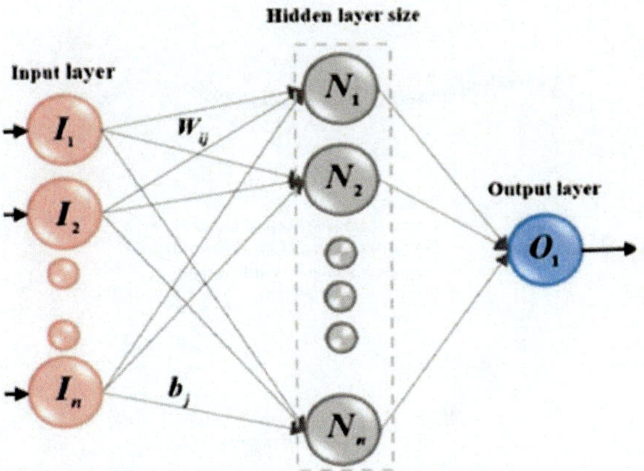

Fig. 1 Architecture of ANN

O_s is the output corresponding to lth data point in the training set by the network,
l_S is the actual output as consider in the target set,
nd is the number of data point considered in training data-set.

4 Numerical Analysis

For the numerical study on the soil-pipeline interaction, the heterogeneity of the soil and the nonlinearity of the pipe have been considered and discretized on 63 elements. The nonlinearity of the pipe and the soil has been taken into account by using the Von Karmàn method to simulate the pipe and the model was validated based on Ref [24] and the ANN method is used based on the collected data from improved FEM to predict the displacement of pipeline. Figure 2 presents the validation of a numerical model with Ref [29]. Pipeline and soil properties are presented in Table 1.

4.1 Results

After validation of our numerical method as done in Ref [24] by comparing the linear and nonlinear response of the pipeline as shown in Fig. 2. For the linear

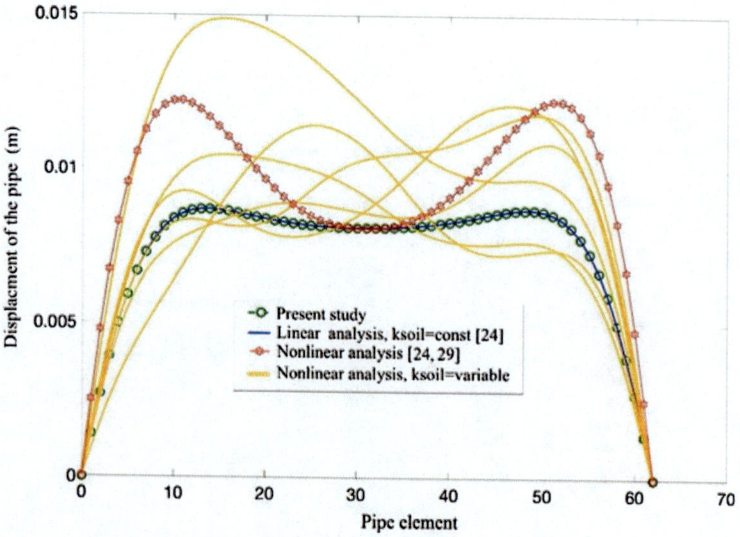

Fig. 2 Linear and nonlinear analysis of pipeline resting elastic soil with k_{soil} = constant and k_{soil} = variable

Table 1 Pipeline and soil properties [24]

Symbol	Parameters	Range of values	Unit
D_{ext}	Diameter of the pipe	1	m
L_p	Length of the pipe	62	m
E_c	Pipe elastic Young's modulus	3×10^4	MPa
$E_c I_c$	Pipe flexural rigidity	869.45	MN/m^2
ν_p	Poisson ration of pipe	0.2	–
kw	Coefficient of Winkler soil's subgrade reaction	3.067	kN/m^3
ν_p	Poisson ratio of soil	0.35	–
E_s	Soil elastic Young's modulus	12.5	MPa

analysis, the coefficient of subgrade reaction of soil is considered constant and for the nonlinear case, the k_{soil} is variable. Therefore, from Fig. 2, we can deduce that the VanMarcke's method combined to Monte Carlo simulations allows us to obtain a higher displacement of the pipe when the value of k_{soil} change (variable). So, it can be noted that the real response is obtained by using the probabilistic approach. The training after collecting the data using ANN is presented in Fig. 3.
Six scenarios are predicted using ANN after training in both cases linear and nonlinear.

Scenario 1 **Linear Problem**
The result of scenario 1 can be found in Fig. 4.

Scenario 2 **Nonlinear Problem**
Figure 5 shows the result of scenario 2 in the case on nonlinear problem.

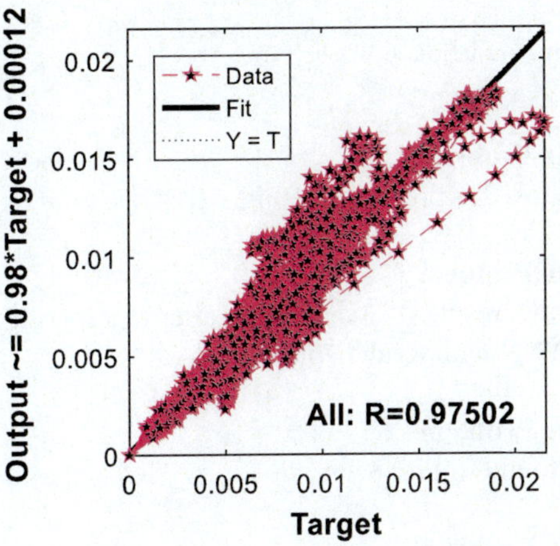

Fig. 3 Training using ANN

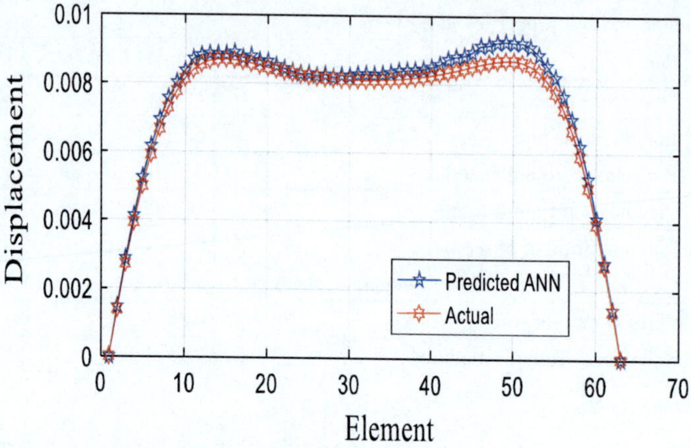

Fig. 4 Scenario 1 predicted displacement using ANN

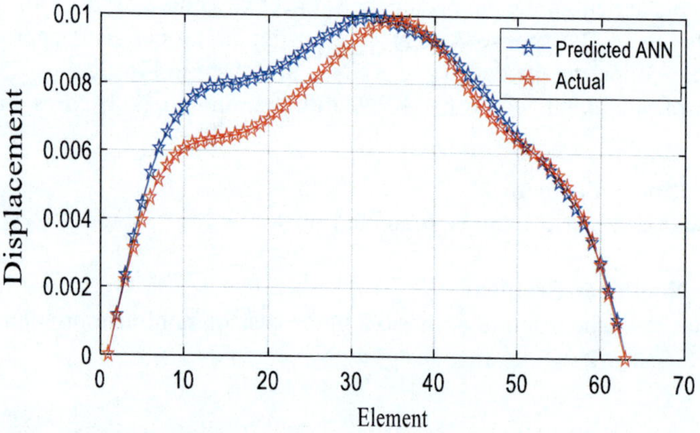

Fig. 5 Nonlinear problem, scenario 2

Scenario 3 **Nonlinear Problem**

The result of scenario 3 is presented in Fig. 6.

Scenario 4 **Nonlinear Problem**

Figure 7 shows the result of scenario 4 after comparing the prediction of displacement using ANN with Real displacement.

Scenario 5 **Nonlinear Problem**

The result can be found in Fig. 8.

Scenario 6 **Nonlinear Problem**

Figure 9 represent the result of scenario 6.

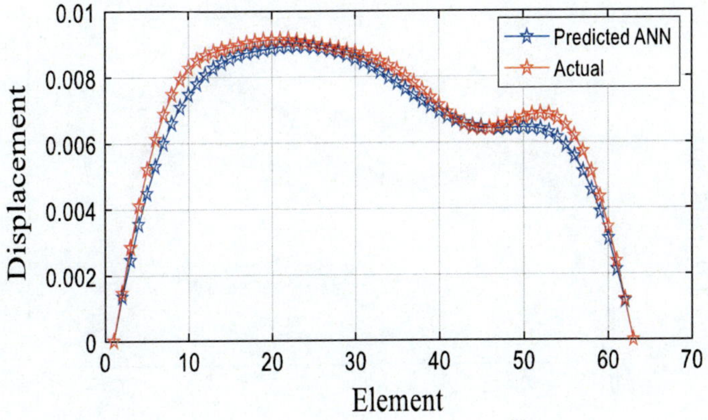

Fig. 6 Nonlinear problem, scenario 3

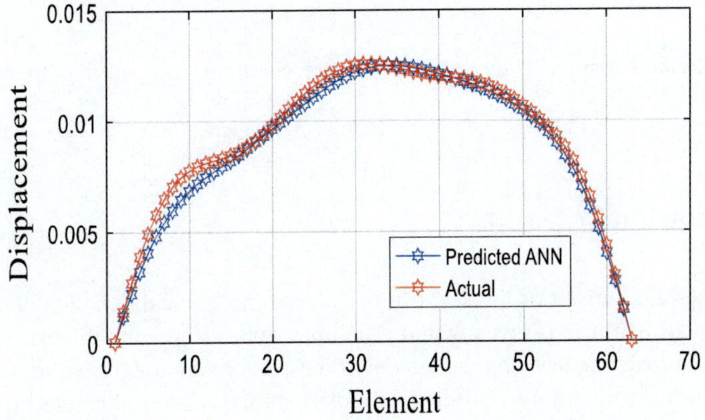

Fig. 7 Nonlinear problem, scenario 4

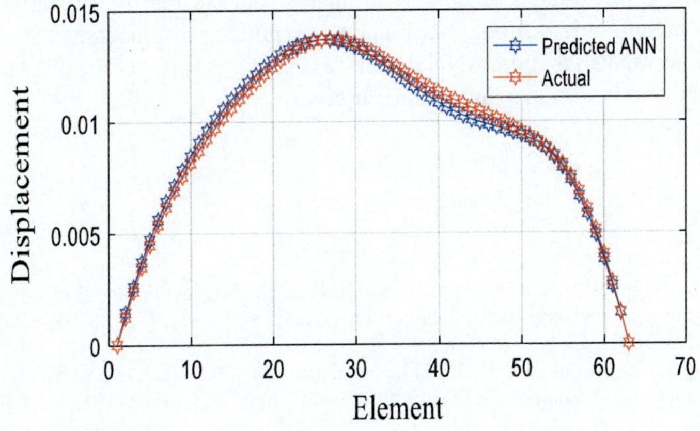

Fig. 8 Nonlinear problem, scenario 5

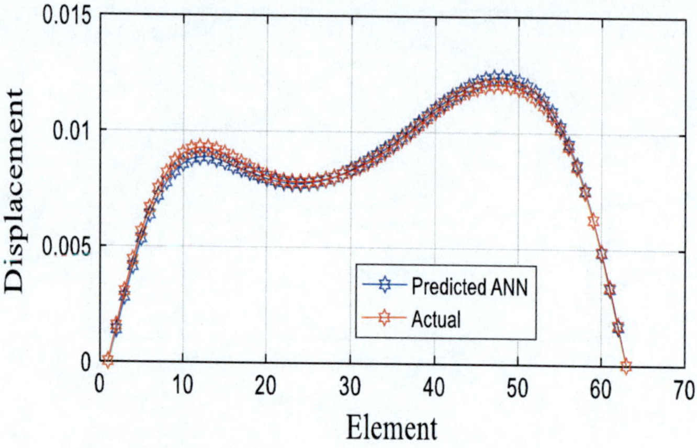

Fig. 9 Nonlinear problem, scenario 6

The provided results using ANN are more accurate compared with reference results.

5 Conclusion

In this paper, the impact of the soil rigidity on the mechanical behaviour for linear and nonlinear pipelines is investigated. The objective is to predict the displacement of pipe using machine learning using ANN. The numerical model is validated based on data from the literature. Different simulations have been generated to obtain data response of the pipe based on displacements. The predicted results using ANN are compared with VanMarcke to prove the effectiveness and the importance of the proposed application. The numerical analysis shows that the variation of the coefficient of subgrade reaction can induce a significant displacement of the pipe. The results prove also that the ANN has more accuracy to predict the displacements of the pipeline in both linear and nonlinear cases.

References

1. Zhang Y, Murphy KD (2004) Response of a finite beam in contact with a tensionless foundation under symmetric and asymmetric loading. Int J Solids Struct 41(24–25):6745–6758. https://doi.org/10.1016/j.ijsolstr.2004.05.028
2. Sapountzakis EJ, Kampitsis AE (2013) Inelastic analysis of beams on two-parameter tensionless elastoplastic foundation. Eng Struct 48:389–401. https://doi.org/10.1016/j.engstruct.(2012).09.012
3. Kuo YH, Lee SY (1994) Deflection of nonuniform beams resting on a nonlinear elastic-foundation. Comput Struct 51(5):513–519. https://doi.org/10.1016/0045-7949(94)90058-2

4. Singh BN, Lal BN, Kumar R (2008) Nonlinear bending response of laminated composite plates on nonlinear elastic foundation with uncertain system properties. Eng Struct 30(4):1101–1112. https://doi.org/10.1016/j.engstruct.(2007).07.007
5. Matsubara K, Hoshiya M (2000) Soil spring constants of buried pipelines for seismic design. J Eng Mech-Asce 126(1):76–83. https://doi.org/10.1061/(asce)0733-9399(2000)126:1(76)
6. Sapountzakis EJ, Kampitsis AE (2013) Nonlinear dynamic analysis of shear deformable beam-columns on nonlinear three-parameter viscoelastic foundation I: theory and numerical implementation. J Eng Mech 139(7):886–896. https://doi.org/10.1061/(asce)em.1943-7889.0000369
7. Kordkheili SAH, Bahai SAH, Mirtaheri M (2011) An updated Lagrangian finite element formulation for large displacement dynamic analysis of three-dimensional flexible riser structures. Ocean Eng 38(5–6):793–803. https://doi.org/10.1016/j.oceaneng.2011.02.001
8. Popescu R, Deodatis G, Nobahar A (2005) Effects of random heterogeneity of soil properties on bearing capacity. Probab Eng Mech 20(4):324–341. https://doi.org/10.1016/j.probengmech.2005.06.003
9. Elachachi SM, Breysse D, Denis A (2021) The effects of soil spatial variability on the reliability of rigid buried pipes. Comput Geotech 43:61–71. https://doi.org/10.1016/j.compgeo.2012.02.008
10. Seguini M, Nedjar D (2017) Nonlinear analysis of deep beam resting on linear and nonlinear random soil. Arab J Sci Eng 42(9) 3875–3893. https://doi.org/10.1007/s13369-017-2449-7
11. Seguini,M., Nedjar, D.: Stochastic finite element analysis of nonlinear beam on Winkler-Pasternack foundation. In: International conference on numerical modelling in engineering, pp 14–29. Springer
12. Grigoriu M, Khater M, O'Rourke T (1985) Stochastic beams on elastic foundation. Struct Safe Stud 96–106: ASCE
13. Griffiths D, Paiboon J, Huang J, Fenton GA (2008) Numerical analysis of beams on random elastic foundations. In: Proceedings of the 9th international congress on numerical methods in engineering and scientific applications, CIMENICS'08, pp 19–25
14. Seguini M, Nedjar D (2016) Nonlinear finite element modeling of deep beam resting on linear and nonlinear random soil. Int J Geol Environ Eng 10(2):134–141
15. Bonnet M, Constantinescu A (2005) Inverse problems in elasticity. Inverse Prob 21(2):R1
16. Kim NH (2014) Introduction to nonlinear finite element analysis. Springer Science & Business Media
17. Nour A, Slimani A, Laouami N, Afra H (2003) Finite element model for the probabilistic seismic response of heterogeneous soil profile. Soil Dyn Earthq Eng 23(5):331–348. https://doi.org/10.1016/s0267-7261(03)00036-8
18. Nedjar D, Hamane M, Bensafi M, Elachachi SM, Breysse D (2007) Seismic response analysis of pipes by a probabilistic approach. Soil Dyn Earthq Eng 27(2):111–115. https://doi.org/10.1016/j.soildyn.2006.06.001
19. Seguini M, Nedjar D (2020) Dynamic and probabilistic analysis of shear deformable pipeline resting on two parameter foundation. Periodica Polytech Civil Eng 64(2):430–437. https://doi.org/10.3311/PPci.14927
20. Seguini M, Boutchicha D, Khatir S, Nedjar D, Le Thanh C, Magd AW (2020) Experimental and numerical vibration analyses of healthy and cracked pipes. In: 1st International conference on structural health monitoring and engineering structures (SHM&ES), pp 15–16
21. Seguini M, Kathir S, Boutchicha D, Nedjar D, Magd AW (2021) Crack prediction in pipeline using ANN-PSO based on numerical and experimental modal analysis. Smart Struct Syst 27(3):507–523. https://doi.org/10.12989/sss.2021.27.3.507
22. Seguini M, Kathir S, Boutchich D, Nedjar D, Magd AW (2021) Crack prediction in beam-like structure using ANN based on frequency analysis. Frattura ed integrità struturalle 16 (59)
23. Khatir S, Magd AW, Benaissa B, Köppen M (2018) Crack identification using extended ISO Geometric analysis and particle swarm optimization. In: Proceedings of the 7th international conference on fracture fatigue and wear. pp 210–222. Springer

24. Seguini M, Nedjar D (2017) Modelling of soil-structure interaction behaviour: geometric nonlinearity of buried structures combined to spatial variability of soil. Eur J Environ Civ Eng 21(10):1217–1236. https://doi.org/10.1080/19648189.2016.1153525
25. Vanmarcke EH (1977) PROBABILISTIC MODELING OF SOIL PROFILES. J Geotech Eng Div-Asce 103(11):1227–1246
26. Vanmarcke E (2010) Random fields: analysis and synthesis. World Sci
27. Fenton GA, Vanmarcke EH (1990) Simulation of random-fields via local average subdivision. J Eng Mech-Asce 116(8):1733–1749. https://doi.org/10.1061/(asce)07339399(1990)116:8(1733)
28. Rukhaiyar S, Alam MN, Samadhiya NK (2018) A PSO-ANN hybrid model for predicting factor of safety of slope. Int J Geotech Eng 12(6):556–566. https://doi.org/10.1080/19386362.2017.1305652
29. Kazi Tani N, Nedjar D, Hamane M (2013) Non-linear analysis of the behaviour of buried structures in random media (vol 17, p 791). Eur J Environ Civil Eng 17(10):1057–1057. https://doi.org/10.1080/19648189.2013.850816

Delamination Detection of Rectangular Laminated Composite Plates by Combining the One-Dimensional and Two-Dimensional Discrete Wavelet Transforms

Morteza Saadatmorad, Ramazan-Ali Jafari-Talookolaei, Mohammad-Hadi Pashaei, Samir Khatir, and Magd Abdel Wahab

Abstract Delamination is one of the most common damages in laminated composite structures. This damage is usually created during manufacturing. Therefore, delamination detection is essential to prevent structural failure in operational conditions. This study proposes a new delamination detection technique by combining the one-dimensional and two-dimensional discrete wavelet transforms. Since delamination is boundary damage, differentiation of its boundaries is significant, but challenging, and the conventional two-dimensional wavelet transformations have weaknesses in overcoming this challenge in some cases. The main idea of the proposed technique is to combine the ability of one-dimensional discrete wavelet transform with two-dimensional discrete wavelet transform to increase the accuracy of delamination detection. Findings show that the proposed technique can significantly improve delamination detection accuracy.

Keywords Delamination detection · Boundary separation technique · Wavelet transforms · Rectangular laminated composite plates

1 Introduction

Laminated composite structures have many applications due to their desired characteristics, such as lightness, high durability, and high strength [1–3]. Because of these favorite characteristics, laminated composite structures have been used in various industries like aerospace engineering, naval engineering, and civil engineering [4–6].

Delaminations as the most common damages in laminated composite structures are an interesting research area in literature [7]. Valdes et al. [8] conducted an investigation on delamination detection in composite laminates from variations of

M. Saadatmorad (✉) · R.-A. Jafari-Talookolaei · M.-H. Pashaei
Department of Mechanical Engineering, Babol Noshirvani University of Technology, Babol, Iran
e-mail: eng.saadatmorad@gmail.com

S. Khatir · M. Abdel Wahab
Soete Laboratory, Faculty of Engineering and Architecture, Ghent University, Ghent, Belgium

M. Abdel Wahab (ed.), *Proceedings of the 10th International Conference on Fracture Fatigue and Wear*, Lecture Notes in Mechanical Engineering,
https://doi.org/10.1007/978-981-19-7808-1_5

their modal characteristics. Swann et al. [9] optimized piezoelectric sensor location for delamination detection in composite laminates. Their objective function was a damage index that compared the voltage signals from an intact composite laminate and a statistically determined delaminated model. The findings showed that the optimum sensor pattern was capable of detecting delaminations in moderately thick composite plates. Shoja et al. [10] detected delamination in composite laminates using low frequency guided waves. The Pearson correlation coefficient was introduced as a criterion for delamination detection. Okabe et al. [11] presented an investigation on delamination detection in composite laminates using dispersion change based on mode conversion of Lamb waves. A novel delamination detection technique was suggested based on the mode conversion. Tong et al. [12] presented an investigation on delamination detection in composite laminates using improved surrogate-assisted optimization.

Some delamination detection based on wavelet transforms in laminated composite structures has been proposed. Yang et al. [13] presented a study titled delamination detection in composite laminate plates using 2D wavelet analysis of modal frequency surface. Modal frequency surface wavelet coefficients were developed as a new indicator for damage localization. Findings showed that damage indicator was highly sensitive to near surface delamination even under noise. A study on computationally efficient delamination detection in composite beams using Haar wavelets was presented by Hein et al. [14]. Detection of delamination damage in composite plates using energy spectrum of structural dynamic responses decomposed by wavelet analysis was presented by Yan et al. [15].

The aim of this work is to propose a new delamination detection of rectangular laminated composite plates by combining the one-dimensional and two-dimensional discrete wavelet transforms.

2 Discrete Wavelet Transforms

Wavelet transform is a signal processing method that can be used to analyze signals and the information in the signals. Wavelet transform or wavelet analysis can be divided into three categories in terms of dimension. In general, there are three types of signals in terms of dimensions: one-dimensional signal, two-dimensional signal, and three-dimensional signal. If one-dimensional signal processing is transformed, one-dimensional wavelet transformation is used, and if two-dimensional and three-dimensional signals are transformed, then two-dimensional and three-dimensional wavelet transformation are used, respectively. Also, in terms of the type of signal analysis, there are two types of wavelet transform: discrete wavelet transform and continuous wavelet transform. By using the one-dimensional discrete wavelet transform (1D-DWT), it is possible to decompose a given 1D-signal $f(x)$ into an approximation signal plus detail signals as: [16]:

$$f(x) = A_j(x) + \sum_{j<J} D_j(x) \tag{1}$$

where j is the level corresponding to the wavelet transform.

The approximation signals corresponded to the level j is stated as:

$$A_j(x) = \sum_{k=-\infty}^{+\infty} cA_{j,k}\phi_{j,k}(x) \tag{2}$$

where $cA_{j,k}$ shows coefficients related to the approximation signal at the level j. In addition, $\phi_{j,k}(x)$ denote the scaling function at the level j.

The detail sub-signals corresponded to the level j are stated as:

$$D_j(x) = \sum_{k \int Z} cD_{j,k}\psi_{j,k}(x) \tag{3}$$

where $cD_{j,k}$ indicate detail coefficients of signal at the level j. In addition, $\psi_{j,k}(x)$ denotes wavelet function.

In wavelet transforms, the vanishing moments are the most significant parameter affecting detecting delaminations or damages in signals. If a wavelet transform has n vanishing moments, then the following relation is satisfied:

$$\int_{-\infty}^{+\infty} x^i \psi(x)\mathrm{d}x = 0, \quad i = 1, 2, \ldots, n-1 \tag{4}$$

The Eq. (1)-(4) can be extended as two-dimensional signals f(x,y) because the 2D-wavelet transformation may be presented in the form of tensor products concerning 1D-wavelet transform as[16]:

$$\phi(\mathrm{x}, \mathrm{y}) = \phi(\mathrm{x})\phi(\mathrm{y})$$

$$\psi^{\mathrm{Horizontal}}(x, y) = \phi(x)\psi(y)$$

$$\psi^{\mathrm{Vertical}}(x, y) = \psi(x)\phi(y) \tag{5}$$

$$\psi^{\mathrm{Diagonal}}(x, y) = \psi(x)\psi(y)$$

where $\phi(x,y)$ is the scaling function. $\psi^H(x, y)$, $\psi^V(x, y)$, and $\psi^D(x, y)$ are the wavelet transformation functions applied in horizontal, vertical, and diagonal directions, respectively.

Therefore, the 2D-signal f(x, y) or the image may be decomposed into four various quarter-sized images (i.e., W_{ϕ} as an approximation sub-image, and W_{ψ}^{H}, W_{ψ}^{V}, W_{ψ}^{D} as detail sub-images in horizontal, vertical, and diagonal directions, respectively) at level j. The relations for the two-dimensional discrete wavelet transform are rewritten as follows:

$$
\begin{aligned}
A\phi_{j,m,n}(x, y) &= \frac{1}{\sqrt{MN}} \sum_{x=0}^{M-1} \sum_{y=0}^{N-1} f(x, y)\phi_{j,m,n}(x, y) \\
D_{j,m,n}^{\text{Horizontal}}(x, y) &= \frac{1}{\sqrt{MN}} \sum_{x=0}^{M-1} \sum_{y=0}^{N-1} f(x, y)\psi_{j,m,n}^{\text{Horizontal}}(x, y) \\
D_{j,m,n}^{\text{Vertical}}(x, y) &= \frac{1}{\sqrt{MN}} \sum_{x=0}^{M-1} \sum_{y=0}^{N-1} f(x, y)\psi_{j,m,n}^{\text{Vertical}}(x, y) \\
D_{j,m,n}^{\text{Diagonal}}(x, y) &= \frac{1}{\sqrt{MN}} \sum_{x=0}^{M-1} \sum_{y=0}^{N-1} f(x, y)\psi_{j,m,n}^{\text{Diagonal}}(x, y)
\end{aligned}
\tag{6}
$$

In this paper, single-level 2D-DWTs decompose the 2D vibration amplitude signals of delaminated rectangular composite plate.

3 The Proposed Method

As mentioned, delamination is one of the most common damages that may occur in laminated composite structures. In this section, the delamination detection in rectangular laminated composite plates is presented using the two-dimensional wavelet method combined with the one-dimensional wavelet method.

The vibration mode shape signal of the delaminated rectangular laminated plates with single rectangular delamination is used to determine the result of damage detection using the two-dimensional discrete wavelet method. In this way, one-dimensional wavelets are applied to the borders to detect the boundaries accurately. The number of 10×10 elements is applied on six-layered plates, and damage scenarios are considered.

As an example, a five-layer composite plate is considered (Fig. 1). According to this figure, the Z-axis is placed in the center of the thickness of the plate, and its positive direction is downwards. According to the figure below, the plates are divided into a limited number of elements, each element has nine nodes, and each node has 5° of freedom.

As seen in Fig. 2, in the proposed delamination detection approach, first mode shape signals are obtained from a finite element model. Then the two-dimensional discrete wavelet transforms are applied to the signal. If the delamination's boundaries

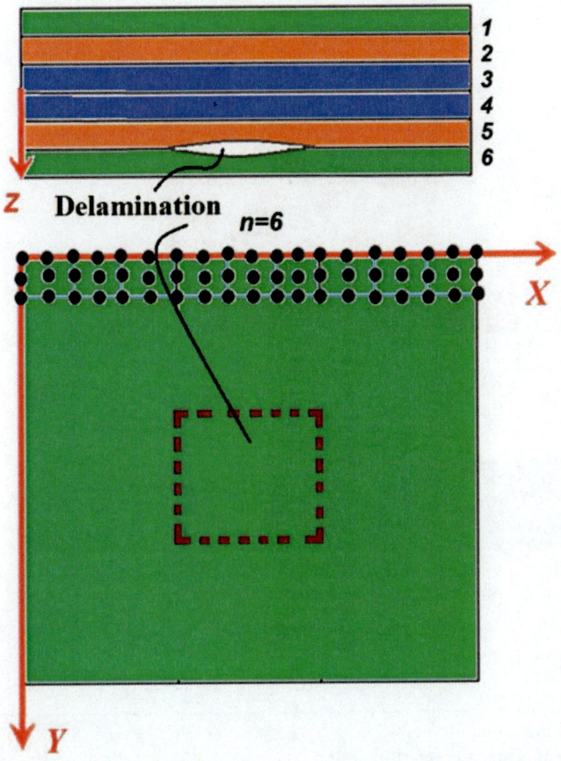

Fig. 1 Interface and location of delamination in the studied six-layered rectangular laminated composite plate

are detected clearly, the delamination detection is performed perfectly. However, if delamination's boundaries are unclear, one-dimensional discrete wavelet transforms are applied on doubtful boundaries.

4 Results

4.1 The First Delamination Scenario

This delamination scenario is described in Table 1. The results of this delamination scenario are shown in Fig. 3. As seen in this figure, the delamination boundaries are well detected in the signal of the diagonal details of the 2D-wavelet coefficients, and there is no need to use a one-dimensional wavelet transform.

The results of the first scenario of delamination detection are shown in Fig. 3. As can be seen, in Fig. 3, there are four signals, three of which are related to details,

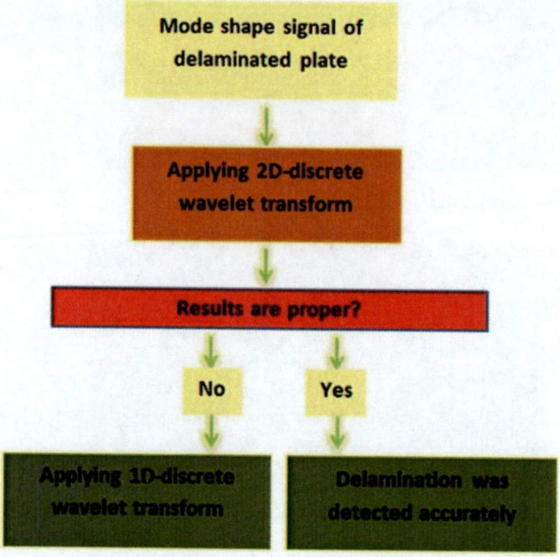

Fig. 2 Flowchart of the proposed delamination detection approach

Table 1 Descriptions of the first delamination scenario

Parameter	Description
Location of delamination in element coordinate m (*results are presented in node coordinate: $n = 2\,m + 1$. For example the plate length 10 in element coordinate is $n = (2 \times 10) + 1 = 21$	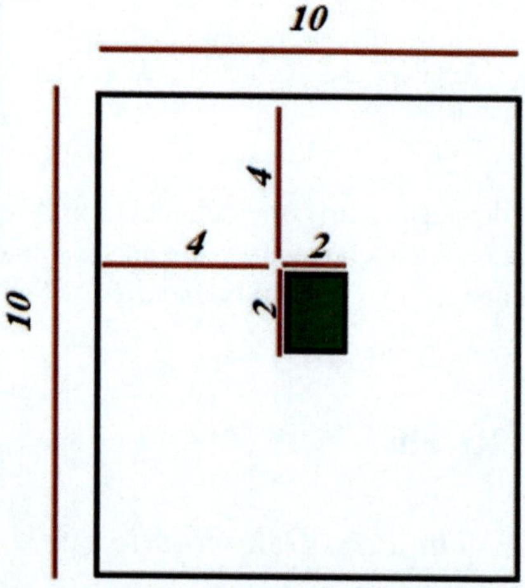
Number of layers of plate	6
Number of elements along X and Y	10, 10
Delaminated interface number	5

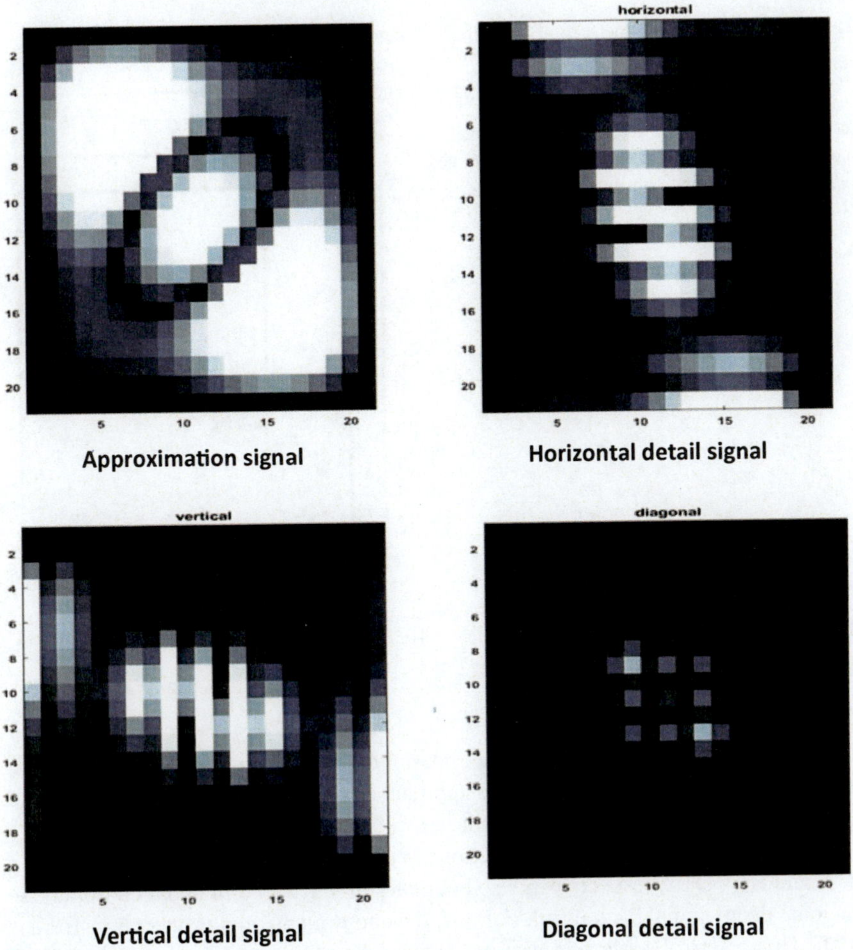

Fig. 3 Results of delamination detection for the first delamination scenario (results are presented in nodal coordinate, i.e., $n = 2\,m + 1$)

and one signal is related to approximation. As seen, the diagonal detail signal of the mode shape can detect the delamination's boundaries well.

4.2 *The Second Delamination Scenario*

The second delamination scenario is described in Table 2. As seen in Fig. 4, there are four signals, three of which are related to details, and one signal is related to approximation. As seen, the diagonal detail signal of the mode shape can detect the delamination's boundaries well. As shown, the delamination is detected in the

Table 2 Descriptions of the second delamination scenario

Parameter	Description
Location of delamination in element coordinate m (*results are presented in node coordinate: $n = 2\,m + 1$. For example the plate length 10 in element coordinate is $n = (2 \times 10) + 1 = 21$	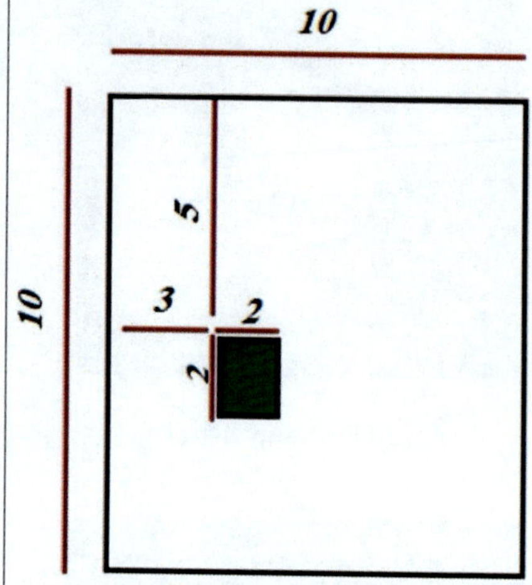
Number of layers of plate	6
Number of elements along X and Y	10, 10
Delaminated interface number	5

diagonal detail signal. However, the boundaries of the delamination are doubtful. For this reason, according to our proposed approach, a one-dimensional discrete wavelet transform is applied to doubtful boundaries.

Maximum values in X and Y directions in diagonal detail signal are named Max D1X and Max D1Y, respectively. The boundary of y at $x = 7$ in nodal coordinate in diagonal detail signal is doubtful. For this, the one-dimensional wavelet transform is applied to $x = 7$ to observe the variations in the Y direction. As demonstrated in Fig. 5, the result of the one-dimensional wavelet transform emphasizes that the location X, $Y = 7$, 15 in nodal coordinate (i.e., X, $Y = 3$, 7) is definitely a boundary point of delamination. The two-dimensional wavelet transform did not emphasize this. Thus, it is demonstrated that the one-dimensional wavelet transform can be used to improve the results of the two-dimensional wavelet transform efficiently.

5 Conclusions

This paper proposes a novel delamination detection technique by combining the abilities of one-dimensional discrete wavelet transform and two-dimensional discrete wavelet transform. According to our findings, in some cases, delamination detection based on the two-dimensional discrete wavelet transform has a weakness for delamination's boundary detection. Thus, to fix this challenge, this paper suggests using the

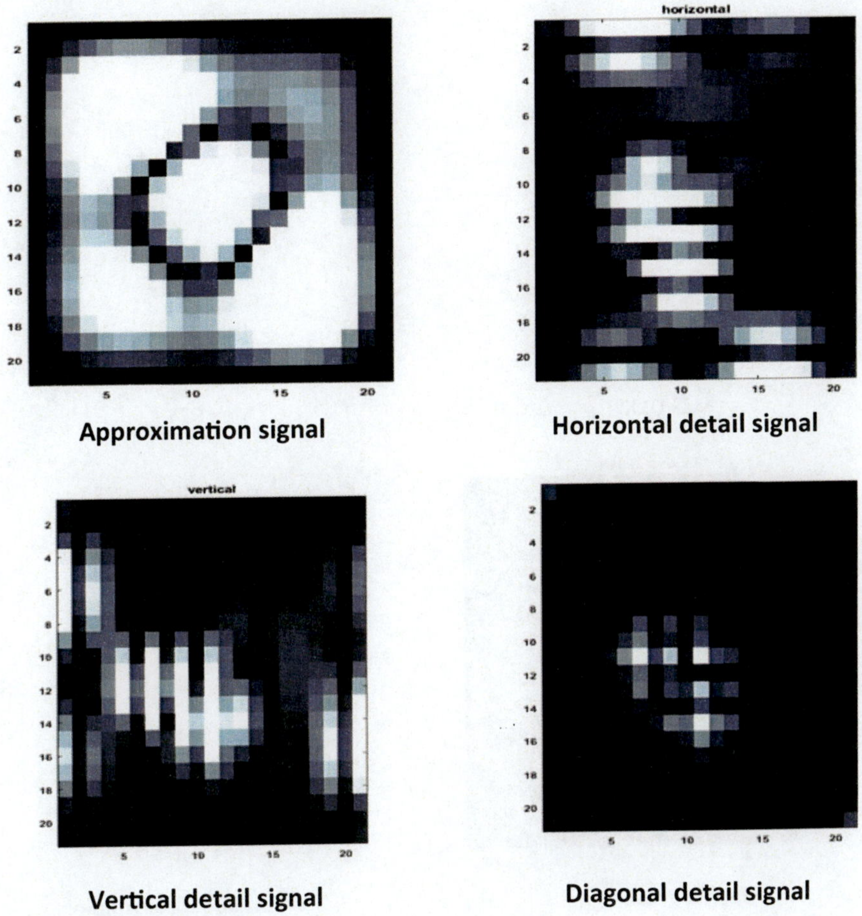

Fig. 4 Results of delamination detection for the second delamination (results are presented in nodal coordinate, i.e., n = 2 m + 1)

one-dimensional discrete wavelet transform in doubtful delamination's boundary. The findings show that the one-dimensional discrete wavelet transform can effectively and efficiently determine the accurate delamination's boundary in doubtful points and be proper support for the two-dimensional discrete wavelet transform and eliminate its weakness.

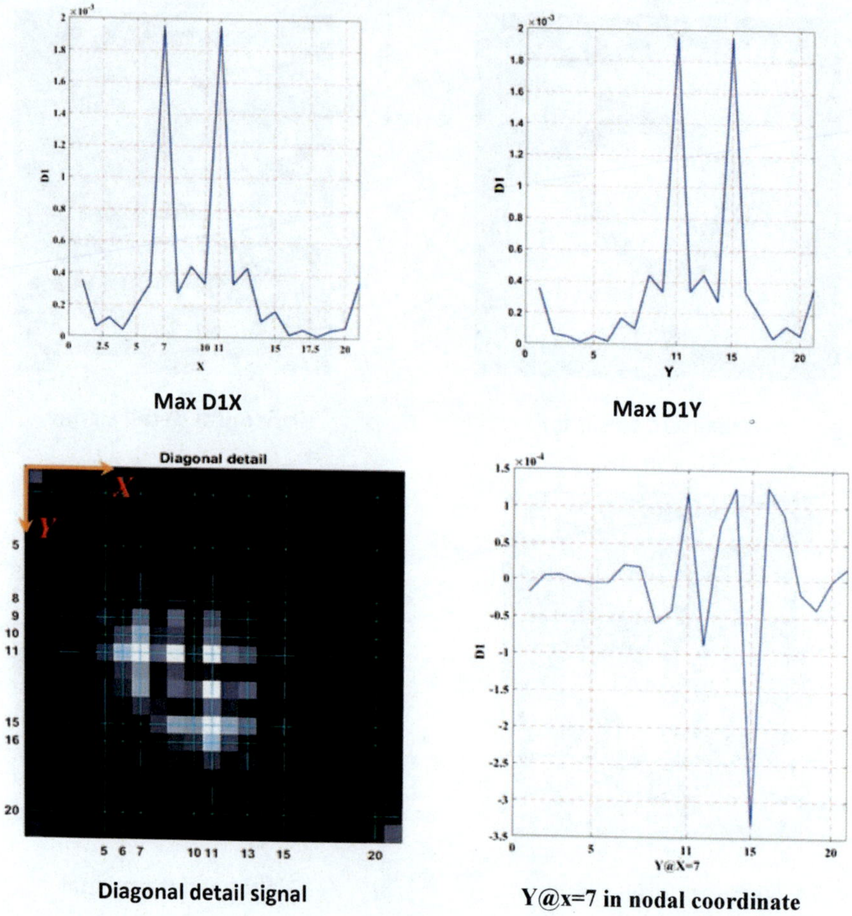

Fig. 5 Results of delamination detection for the second delamination after applying one-dimensional wavelet transform (results are presented in nodal coordinate, i.e., $n = 2m + 1$)

References

1. He K, Hoa SV, Ganesan R (2000) The study of tapered laminated composite structures: a review. Compos Sci Technol 60(14):2643–2657
2. Nikbakt S, Kamarian S, Shakeri M (2018) A review on optimization of composite structures part I: laminated composites. Compos Struct 195:158–185
3. Karama M, Afaq KS, Mistou S (2003) Mechanical behaviour of laminated composite beam by the new multi-layered laminated composite structures model with transverse shear stress continuity. Int J Solids Struct 40(6):1525–1546
4. Williams KV, Vaziri R, Poursartip A (2003) A physically based continuum damage mechanics model for thin laminated composite structures. Int J Solids Struct 40(9):2267–2300
5. Sridharan S (ed) (2008) Delamination behaviour of composites. Elsevier

6. Mi Y, Crisfield MA, Davies GAO, Hellweg HB (1998) Progressive delamination using interface elements. J Compos Mater 32(14):1246–1272
7. Iarve EV, Gurvich MR, Mollenhauer DH, Rose CA, Dávila CG (2011) Mesh-independent matrix cracking and delamination modeling in laminated composites. Int J Numer Meth Eng 88(8):749–773
8. Valdes SD, Soutis C (1999) Delamination detection in composite laminates from variations of their modal characteristics. J Sound Vib 228(1):1–9
9. Swann C, Chattopadhyay A (2006) Optimization of piezoelectric sensor location for delamination detection in composite laminates. Eng Optim 38(5):511–528
10. Shoja S, Berbyuk V, Boström A (2018) Delamination detection in composite laminates using low frequency guided waves: numerical simulations. Compos Struct 203:826–834
11. Okabe Y, Fujibayashi K, Shimazaki M, Soejima H, Ogisu T (2010) Delamination detection in composite laminates using dispersion change based on mode conversion of lamb waves. Smart Mater Struct 19(11):115013
12. Tong H, Pan J, Singh HK, Luo W, Zhang Z, Hui D (2021) Delamination detection in composite laminates using improved surrogate-assisted optimization. Compos Struct 277:114622
13. Yang C, Oyadiji SO (2017) Delamination detection in composite laminate plates using 2D wavelet analysis of modal frequency surface. Comput Struct 179:109–126
14. Hein H, Feklistova L (2011) Computationally efficient delamination detection in composite beams using Haar wavelets. Mech Syst Signal Process 25(6):2257–2270
15. Yan YJ, Yam LH (2004) Detection of delamination damage in composite plates using energy spectrum of structural dynamic responses decomposed by wavelet analysis. Comput Struct 82(4–5):347–358
16. Saadatmorad M, Jafari-Talookolaei RA, Pashaei MH, Khatir S (2021) Damage detection on rectangular laminated composite plates using wavelet based convolutional neural network technique. Compos Struct 278:114656

A Feasibility Review of Novel Avian-Based Optimization Algorithms for Damage Detection in a Truss Bridge

Lan Ngoc-Nguyen, Samir Khatir, Hoa Ngoc-Tran, Huu-Quyet Nguyen, Long Ngoc-Nguyen, Thanh Bui-Tien, and Magd Abdel Wahab

Abstract During their lifecycle, bridge structures have to withstand various uncertainties loads such as wind, typhoon, and accident loads which may pose serious threats to the integrity as well as the safety of the structure, especially when they induced significant damages to the structure. For many years, researchers have been trying to develop heath monitoring tools, which can identify accurately not only the location, but also the level of structural damage. In this paper, two novel avian-based optimization algorithms-Artificial Hummingbird Algorithm (AHA) and African Vulture Optimization Algorithm (AVOA) are reviewed for their feasibility in detecting structural damages in truss bridge. The accuracy of the proposed algorithms is compared against two other famous algorithms: particle swarm optimization (PSO) and cuckoo search (CS). The results of the feasibility review for damage detection capability are discussed.

Keywords Structural health monitoring · Artificial Hummingbird Algorithm · African Vulture Optimization Algorithm · Damage detection

1 Introduction

During their lifecycle, bridge structures have to withstand various uncertain loads such as wind, typhoons, and accident loads which may pose intimidating threats to the structure's integrity and safety. For years, many efforts have been spent on researching the "ultimate" damage detection toolbox, which has the ability to correctly identify both the location and the level of the structural damage. The toolbox would not only help to reduce the overall maintenance cost of the structure but also prolong

---

L. Ngoc-Nguyen (✉) · S. Khatir · H. Ngoc-Tran · M. Abdel Wahab
Soete Laboratory, Faculty of Engineering and Architecture, Ghent University, Ghent, Belgium
e-mail: ngoclan.nguyen@ugent.be

L. Ngoc-Nguyen · H. Ngoc-Tran · H.-Q. Nguyen · L. Ngoc-Nguyen · T. Bui-Tien
Faculty of Civil Engineering, University of Transport and Communications, Hanoi, Vietnam

M. Abdel Wahab (ed.), *Proceedings of the 10th International Conference on Fracture Fatigue and Wear*, Lecture Notes in Mechanical Engineering,
https://doi.org/10.1007/978-981-19-7808-1_6

its service life. While the "holy grail" toolbox has yet to be found, researchers have obtained encouraging results in solving damage detection problems in structure. For example, data-based techniques have been applied to detect structural damages in bridges. Bui-Tien et al. [1] proposed a hybrid method combining recurrent neural network (RNN) and convolutional neural network (CNN) to filter damages in the Z-24 bridge in Switzerland. The method was able to detect structural damages in different scenarios with a high level of accuracy. Neves et al. [2] proposed an ANN-based approach to identify damages in a railway bridge. Corbally and Malekjafarian [3] introduced an improved ANN-based method using the contact-point (CP)-response of passing vehicles to detect structural damages on a simulated bridge. The proposed method showed the superior result in damage detection's capability in comparison to the traditional methods. Ref. [4] introduced a new data-driven damage detection tool combining train-borne signals and deep learning to identify damage to a numerical railway bridge model, which was able to detect damages in different generated scenarios with a high level of accuracy. While data-driven methods usually prove to be effective, their complexity and computational cost can make them less attractive when dealing with small and medium size structures in comparison to optimization-based methods.

In the past decades, optimization techniques have become more and more significant in solving convoluted problems of our everyday life, from the scale of microstructures to lifeline megastructures. Optimization techniques are our key to sustainable development as they enable us to get access to efficient resource management. Traditionally, there are two approaches to optimization techniques: deterministic and meta-heuristic with the latter one has been a hot subject of research recently due to its simplicity, arbitrary, and robustness [5]. With its advantages, many meta-heuristic algorithms have been developed lately [6–11]. Some researchers have also tried to use metaheuristic algorithms to solve damage detection problems of bridges explicitly [12–19].

In this paper, two novel avian-based optimization algorithms—Artificial Hummingbird Algorithm (AHA) and African Vulture Optimization Algorithm (AVOA) are reviewed for their feasibility in solving damage detection of a real-life truss bridge. Their accuracy and effectiveness are also compared against two others widely used optimization algorithms: Cuckoo Search (CS) and Particle Swarm Optimization (PSO); both are also inspired by the avian behaviour of the animal kingdom. The results are then discussed in the conclusion part of the paper.

2 Novel Avian-Based Algorithms

2.1 Artificial Hummingbird Algorithm

The artificial Hummingbird Algorithm was introduced by Zhao et al. [20] in 2022. The algorithm is inspired by the foraging behaviour of the hummingbird, which is one of the smallest kinds of birds in the world but also is amongst the smartest of the animal kingdom. Hummingbirds have an extraordinary memory; they are able to memorize the exact location and food quantity of the previously located food source. This enables them to forage food with a high level of sustainability. AHA mimics three strategies of hummingbirds during their foraging process: guided foraging, territorial foraging, and migration foraging. During their forage, all three strategies shall be conducted to ensure that the hummingbird will always locate and head to the best possible food source nearby.

Strategy 1: Guided foraging

The guided foraging process of AHA mimicked the hummingbird's ability to search for the food source with the highest quantity of food $v_i(t)$. This can be done simply by looking for the food source that it has not visited again for the longest amount of time, since the longer the bird has not visited the food source, the higher the chance the food source will refill itself. The guided foraging strategy is shown in Eq. (1) below:

$$a \cdot D \cdot \left(x_i(t) - x_{i,\text{tar}}(t)\right) + x_{i,\text{tar}}(t) = v_i(t+1) \tag{1}$$

where $x_i(t)$ is the position of the ith food source at time t, $x_{i,\text{tar}}(t)$ is the position of the targeted food source that the ith hummingbird tends to fly to. a and D are the guiding factor and flight direction factor of the bird, respectively

Strategy 2: Territorial foraging

Once the food source is finished, a hummingbird will fly to a nearby region containing a new food source instead of flying back to the older ones. This gives time for the food source to refill itself slowly while also enabling the bird to find a better food source at the same time. The territorial foraging strategy is indicated as follows:

$$b \cdot D \cdot x_i(t) v_i + x_i(t) = v_i(t+1) \tag{2}$$

where $x_i(t)$ is the position of the ith food source at time t, b and D are territorial factors the flight direction factor of the hummingbird.

Strategy 3: Migration foraging

When the food sources within a particular region are diminished, a hummingbird will migrate to a nearby region to look for food. The migration foraging process is shown in Eq. (3) as below:

$$LB + r \cdot (UB - LB) = x_{\text{diminished}}(t+1) \tag{3}$$

where $x_{\text{diminished}}$ is the nearly depleted food source in the population, LB and UB are the lower boundaries and upper boundaries of the search.

2.2 African Vulture Optimization Algorithm

First introduced in 2021 by Abdollahzadeh et al. [21], the African Vulture optimization algorithm is inspired by the behaviour of African vultures in a group, especially during hunting. African vultures are hunters by group; their hunting strategies involve their endurance to fly for hours to look for food to help them survive from starving. The algorithm consists of four main steps: determining the best vulture, calculating the starvation rate amongst the group, exploration, and exploitation.

Step 1: Determining the best vulture

After initialization, the vultures are divided into two groups: the best group and the second-best one. The best solution is selected as the best vultures in each group. The selection is shown in Eq. (4) below:

$$R(i) = \begin{cases} \text{Best Vulture}_1 \text{ if } p_i = L_1 \\ \text{Best Vulture}_1 \text{ if } p_i = L_2 \end{cases} \tag{4}$$

where $R(i)$ is the best vulture selected, L_1 and L_2 are random parameters with values between 0 and 1.

Step 2: Calculating starvation rate amongst the group

Vultures will go hunting for food when hungry, and the hunger drives them to fly long for hours to search for food. However, the hunger also prevents them from having enough energy for a long flight and makes them aggressive. The rate of starvation amongst the group of vultures is indicated as below:

$$F = (2 \times \text{rand}_1 + 1) \times z \times \left(1 - \frac{\text{intertion}_i}{\text{maxiterations}}\right) + t \tag{5}$$

where F indicates the satiation of the vulture, z is a random parameter with a value between 0 and 1, t is the starvation rate factor.

Step 3: Exploration

In the AVOA, African vultures are assumed to acquire different strategies when searching for food in other areas. The parameter P_1 is selected to represent either the strategy and has a value between 0 and 1. Equation (6) below indicates the exploration step of the AVOA:

$$P(i+1) = \begin{cases} R(i) - |X \times R(i) - P(i)| \times F & \text{if } P_1 \geq \text{rand}_{P_1} \\ R(i) - F + \text{rand}_2 \times ((ub - lb) \times \text{rand}_3 + lb) & \text{if } P_1 < \text{rand}_{P_1} \end{cases} \tag{6}$$

where $P(i + 1)$ is the position of the vulture in the next iteration, X is the coefficient factor that indicates the movement of the vulture to protect the food from the others.

When $0.5 < |F| < 1$, the first exploitation phase is conducted. In the first phase of exploitation, two siege-fight strategies of vultures are assumed when they content for food. A parameter P_2 is selected to represent either the strategy and has a value between 0 and 1. The first phase of exploitation is indicated as:

$$P(i+1) = \begin{cases} |X \times R(i) - P(i)| - (F + \text{rand}_4) - R(i) + P(i) & \text{if } P_2 \geq \text{rand}_{P_2} \\ R(i) - (S_1 + S_2) & \text{if } P_2 < \text{rand}_{P_2} \end{cases} \tag{7}$$

where S1, S2 are the spiral rotational flight of vultures [21].

When $|F| < 0.5$, the second phase is performed. The different movement within the group of vultures paved way for the two aggressive siege fight among the vultures. A parameter P_3 is selected to represent either the strategy and has a value between 0 and 1. The second phase is shown in Eq. (8) below:

$$P(i+1) = \begin{cases} 0.5(A_1 + A_2) & \text{if } P_3 \geq \text{rand}_{P_3} \\ R(i) - |R(i) - P(i)| \times F \times \text{Levy}(d) & \text{if } P_3 < \text{rand}_{P_3} \end{cases} \tag{8}$$

where A_i is the movement of the vulture towards the food source [21], Levy(d) is the levy flight [22] for the distance of the vulture to the best vulture of the respective group.

3 Case Study: Damage Detection of Nam O Bridge

3.1 Introduction

In this section, the Nam O truss bridge is chosen as the real-life case study to review the feasibility of the two mentioned-above algorithms in solving the damage detection problem of structure. The Nam O Bridge (Fig. 1) is a railway truss arch bridge located in the middle of Vietnam. The bridge holds a strategic position as it connects the busy North–South Railway Line of the country. The truss arch bridge consists of four truss spans of 75 m each. Roller and pin bearings are used to support the bridge span. The truss members include lower chords, upper chords, vertical chords, diagonal chords, struts, lower, and upper wind bracings. Truss members are connected to each other by truss bolts.

Fig. 1 Nam O bridge with a regular train passed by

3.2 Baseline Finite Element Model

A Finite Element (FE) model of the first bridge span is modelled using Stabil toolbox of MATLAB [23]. The model includes a total of 45 nodes and 146 elements with six degrees of freedom (DOF) at default at each node: translational displacements in X, Y, Z axes and rotational displacement at the respective axes. In this model, all non-structural components such as parapets, maintenance paths, etc., are considered as added concentrated loads. The FE model of the bridge is shown in Fig. 2 as below:

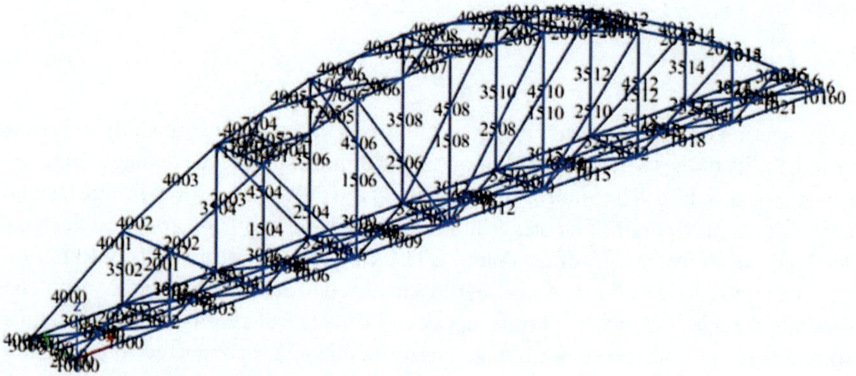

Fig. 2 FE model of the truss arch span

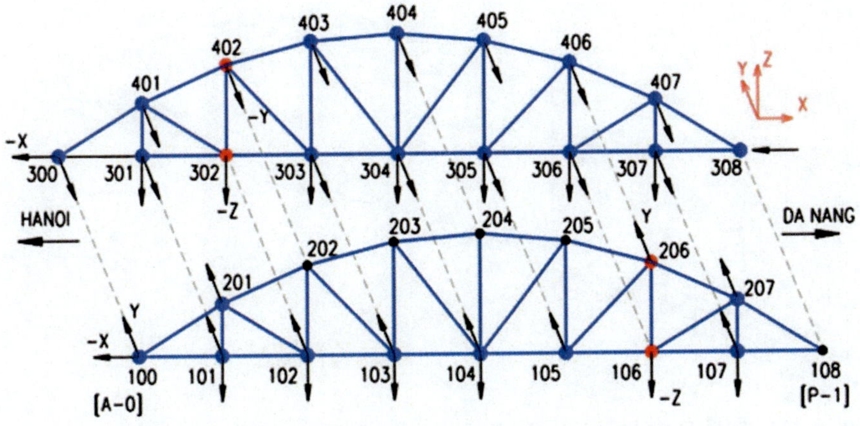

Fig. 3 Measurement grid. Blue: roving points; red: reference points [14]

A vibration measurement campaign of the bridge was conducted to analyse the dynamic characteristic of the actual bridge. Since Nam O is a railway bridge, the measurement considered two sources of excitation: ambient vibration from the wind and the free vibration of the bridge after the train has passed by. The data acquisition system consists of eight high sensitivity accelerometers of type PCB393B12 which were attached to the truss member by magnets, an 8 slot Compact DAQ Chassis (cDAQ 9178) connected to two NI-9234 modules, and a laptop installed with signal processing program MACEC [24]. A measurement grid was set up (Fig. 3) which covered all vibration sensitive location along the measured span.

Measurement data are analyzed by MACEC to obtain the natural frequencies and the mode shapes of the bridge. Once the measured dynamic characteristics are obtained, model updating is performed to get a baseline FE model of the bridge, which will be used for the damage detection process in the next section. Details of the model updating are explained briefly in [14]. The result of model updating for the first 05 vibration modes are shown in Table 1.

Table 1 Natural frequencies before and after model updating

No.	Measured frequency (Hz)	Initial FE model's frequency (Hz)	Updated FE model's frequency (Hz)
1	1.45	1.18 (18.6%)	1.47 (1.4%)
2	3.11	2.76 (11.3%)	3.14 (1.0%)
3	3.28	3.11 (5.2%)	3.32 (1.2%)
4	4.62	3.79 (17.7%)	4.80 (3.7%)
5	6.05	3.94 (34.9%)	6.96 (13.0%)

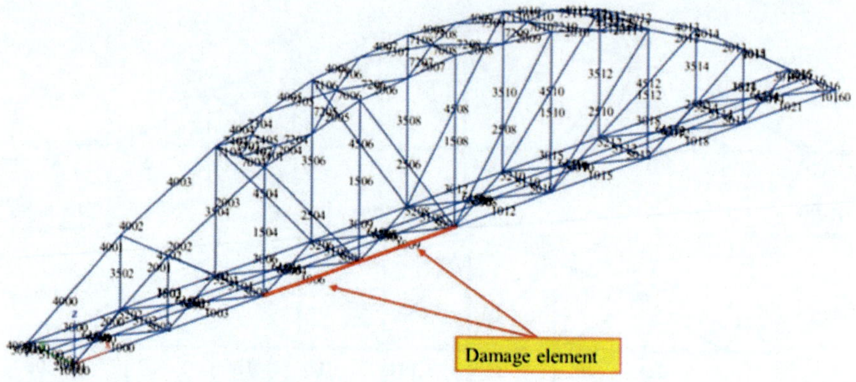

Fig. 4 Damage elements of Nam O bridge

3.3 *Damage Detection Using Proposed Algorithms*

In this section, to study the feasibility for damage detection of the two proposed method, two damage scenarios are generated based on the simulated model. In the first scenario, single damage is introduced by reducing the Young's modulus of element number 1006 at the bottom chord to 30%. In the second scenarios, multiple damages are generated by reducing the Young's modulus of element number 1006 and 1009 at the bottom chord to 30% and 50% respectively (Fig. 4).

The objective function for the two case is as below:

$$\text{Objective Function} = \sum_{i=1}^{5} \left(f_i - \tilde{f}_i\right)^2 / \tilde{f}_i^2 \tag{9}$$

where f_i, $\tilde{f}_i$ are the natural frequencies of the analytical and experimental values, respectively, for the first 05 vibration modes of the structure. To evaluate the effectiveness of the proposed methods, they are compared with two other well known avian-inspired algorithms: Cuckoo Search (CS) and Particle Swarm Optimization (PSO). In all of the algorithms, 50 population size and 50 iterations are selected for each run. The results are shown as below:

For the single damage case:

For the multiple damage case:

From Figs. 5 and 6, we can see that in both single and multiple damage scenarios, AHA and AVOA have a better convergence rate than CS and PSO. AHA and AVOA are able to identify correctly not only the location of damage but also the level of damage. For the single damage case, AHA and AVOA located accurately the damage at the first element, while CS and PSO provided false damage at element number 4, 6, and 10, respectively. For the multiple damage case, the same thing happened with false damages appeared at the incorrect location in the case of CS and PSO.

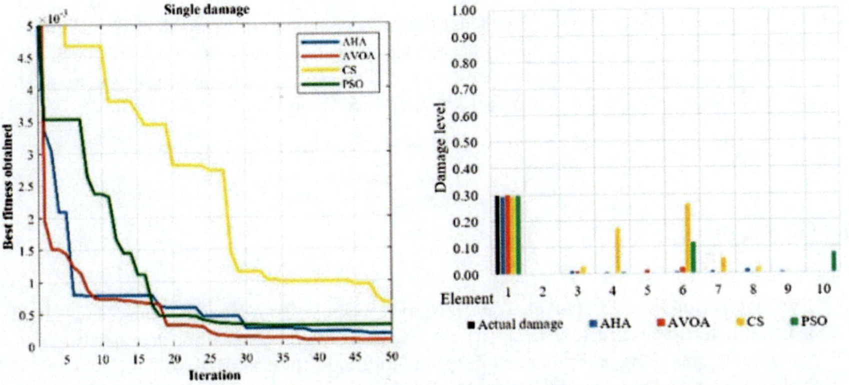

Fig. 5 Convergence of fitness function (left) and single damage detection of Nam O bridge (right)

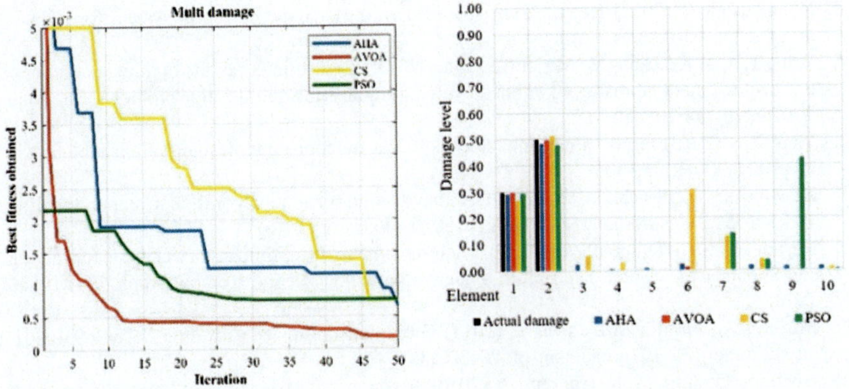

Fig. 6 Convergence of fitness function (left) and multiple damage detection of Nam O bridge (right)

4 Conclusion

In this paper, a feasibility review of the damage detection capability of two novel avian-based optimization algorithm—AHA and AOVA is conducted. The damage detection capability is validated on a real-life truss arch bridge structure, with different damage scenarios are generated by adjusting the stiffness of the respective element. The proposed algorithms are compared with two other older avian-inspired algorithm—CS and PSO for accuracy and effectiveness. The analysed results show that the proposed algorithms have been able to localize and quantify structural damages with a high level of accuracy. The two algorithms are effective and feasible in dealing with structural damage detection problems. In the future, since both algorithms are recently developed, then can be further improved to deal with more complex engineering problems such as damage detection of large-scale bridges, skyscrapers, etc.

Acknowledgements The authors acknowledge the financial support of VLIR-UOS TEAM Project, VN2018TEA479A103, 'Damage assessment tools for Structural Health Monitoring of Vietnamese infrastructures', funded by the Flemish Government. This work is also funded by Vingroup and supported by Innovation Foundation (VINIF) under project code VINIF.2021.DA00192.

References

1. Bui-Tien T, Bui-Ngoc D, Nguyen-Tran H, Nguyen-Ngoc L, Tran-Ngoc H, Tran-Viet H (2022) Damage detection in structural health monitoring using hybrid convolution neural network and recurrent neural network. Frattura ed Integrità Strutturale 16(59):461–470
2. Neves AC, Gonzalez I, Leander J, Karoumi R (2017) Structural health monitoring of bridges: a model-free ANN-based approach to damage detection. J Civ Struct Health Monit 7(5):689–702
3. Corbally R, Malekjafarian A (2022) A data-driven approach for drive-by damage detection in bridges considering the influence of temperature change. Eng Struct 253:113783
4. Hajializadeh D (2022) Deep-learning-based drive-by damage detection system for railway bridges. Infrastructures 7(6):84
5. Wang L, Cao Q, Zhang Z, Mirjalili S, Zhao W (2022) Artificial rabbits optimization: a new bio-inspired meta-heuristic algorithm for solving engineering optimization problems. Eng Appl Artif Intell 114:105082
6. Yang X-S (2010) Firefly algorithm, stochastic test functions and design optimisation. arXiv preprint arXiv:1003.1409
7. Zervoudakis K, Tsafarakis S (2020) A mayfly optimization algorithm. Comput Ind Eng 145:106559. https://doi.org/10.1016/j.cie.2020.106559
8. Mirjalili S (2016) Dragonfly algorithm: a new meta-heuristic optimization technique for solving single-objective, discrete, and multi-objective problems. Neural Comput Appl 27(4):1053–1073
9. Mirjalili S, Mirjalili SM, Lewis A (2014) Grey wolf optimizer. Adv Eng Softw 69:46–61. https://doi.org/10.1016/j.advengsoft.2013.12.007
10. Abedinia O, Amjady N, Ghasemi A (2016) A new metaheuristic algorithm based on shark smell optimization. Complexity 21(5):97–116. https://doi.org/10.1002/cplx.21634
11. Abualigah L, Shehab M, Alshinwan M, Mirjalili S, Elaziz MA (2021) Ant lion optimizer: a comprehensive survey of its variants and applications. Arch Comput Methods Eng 28(3):1397–1416
12. Nguyen-Ngoc L, Tran-Ngoc H, Nguyen-Tran H, Nguyen-Duc B, Nguyen-Le-Minh D, Bui-Tien T, Abdel Wahab M (2021) Damaged detection in structures using artificial neural networks and genetic algorithms. In: Bui-Tien T, Nguyen Ngoc L, De Roeck G (eds) Proceedings of the 3rd international conference on sustainability in civil engineering. Springer, Singapore, pp 33–38
13. Bui-Tien T, Quang NT, Ho LV (2021) A hybrid heuristic optimization algorithm PSOGSA coupled with a hybrid objective function using ECOMAC and frequency in damage detection, p 15
14. Tran-Ngoc H, Khatir S, De Roeck G, Bui-Tien T, Nguyen-Ngoc L (2018) Abdel WM model updating for Nam o bridge using particle swarm optimization algorithm and genetic algorithm. Sensors 18(12):4131
15. Nguyen-Ngoc L, Tran NH, Bui-Tien T, Mai-Duc A, Abdel Wahab M, Nguyen HX, De Roeck G (2021) Damage detection in structures using particle swarm optimization combined with artificial neural network. Smart Struct Syst 28(1):1–12
16. He X, Kawatani M, Hayashikawa T, Furuta H, Matsumoto T (2011) A bridge damage detection approach using train-bridge interaction analysis and GA optimization. Proc Eng 14:769–776
17. Mohammed YM, Uddin N (2017) Bridge damage detection using the inverse dynamics optimization algorithm. In: 26th ASNT research symposium, pp 175–184

18. Tran-Ngoc H, Khatir S, De Roeck G, Bui-Tien T, Wahab MA (2019) An efficient artificial neural network for damage detection in bridges and beam-like structures by improving training parameters using cuckoo search algorithm. Eng Struct 199:109637
19. Gomes GF, Mendez YAD, da Silva Lopes Alexandrino P, da Cunha SS, Ancelotti AC (2019) A review of vibration based inverse methods for damage detection and identification in mechanical structures using optimization algorithms and ANN. Arch Comput Methods Eng 26(4):883–897
20. Zhao W, Wang L, Mirjalili S (2022) Artificial hummingbird algorithm: a new bio-inspired optimizer with its engineering applications. Comput Methods Appl Mech Eng 388:114194
21. Abdollahzadeh B, Gharehchopogh FS, Mirjalili S (2021) African vultures optimization algorithm: a new nature-inspired metaheuristic algorithm for global optimization problems. Comput Ind Eng 158:107408
22. Yang XS (2010) Nature-inspired metaheuristic algorithms. Luniver Press
23. Dooms D, Jansen M, De Roeck G, Degrande G, Lombaert G, Schevenels M, François S (2010) StaBIL: a finite element toolbox for MATLAB. Version 2.0 User's Guide
24. Reynders E, Schevenels M, De Roeck G (2014) MACEC 3.3. A Matlab toolbox for experimental and operational modal analysis. User Manual, Leuven

Static Assessment of Notched Additively Manufactured Polymers Based on the Theory of Critical Distances

Luca Susmel

Abstract The aim of this paper is to review the work we have done in recent years to investigate the accuracy of the Theory of Critical Distances (TCD) in estimating static strength of notched additively manufactured acrylonitrile butadiene styrene (ABS) and polylactide (PLA). The TCD takes as its starting point the assumption that the extent of damage under static loading can be assessed successfully by using two different material parameters, i.e. (i) a critical distance whose length is closely related to the material micro/meso/macro-structural features and an inherent (i.e. a defect free) material strength. Plain and notched specimens of 3D-printed PLA and FDM were manufactured (with an in-fill level of 100%) by making the deposition angle vary in the range 0°–90°. Using the TCD, failures were predicted by directly postprocessing the linear-elastic stress fields determined by solving standard linear-elastic Finite Element (FE) models. Independently of notch sharpness and printing direction, the estimates being obtained were found to be highly accurate, falling within an error interval of about 20%. This result fully supports the idea that the TCD can successfully be used in situations of practical interest to design against static loading notched components of additively manufactured polymers by directly postprocessing the results from simple linear-elastic FE models.

Keywords Additive manufacturing · PLA · FDM · Notch · Critical distance

1 Introduction

According to ASTM F42, additive manufacturing (AM) is "*the process of joining materials to make objects from 3D-model data, usually layer upon layer, as opposed to subtractive manufacturing methodologies*". Thanks to its specific features, AM allows complex shapes to easily be incorporated in structural components.

L. Susmel (✉)
Department of Civil and Structural Engineering, The University of Sheffield, Sheffield S1 3JD, UK
e-mail: l.susmel@sheffield.ac.uk

M. Abdel Wahab (ed.), *Proceedings of the 10th International Conference on Fracture Fatigue and Wear*, Lecture Notes in Mechanical Engineering,
https://doi.org/10.1007/978-981-19-7808-1_7

If attention is focussed on thermoplastic polymers, certainly acrylonitrile butadiene styrene (ABS) and polylactide (PLA) are the most common materials that are used to additively manufacture objects at a relatively low cost. Polymers can be additively manufactured (AM) from powders, wires and flat sheets that are melted using a variety of different technological processes. PLA is a biodegradable, absorbable and biocompatible polymer that is widely employed to manufacture biomedical components. Owing to its specific mechanical response and resistance to corrosion, ABS is commonly used to manufacture rigid and lightweight components. PLA and ABS can easily be additively manufactured by using low-cost commercial 3D-printers. Other than these two materials, polyphenyl sulfone and polycarbonate as well can be 3D-printed effectively, although they can be used provided that more advanced AM techniques are used.

One of the most relevant features of AM is that objects can be fabricated in a very accurate way, with this holding true in terms of both shape and dimensions. As far as structural integrity is concerned, the fact that 3D-printed components can contain geometrical features of all kinds results in localised stress concentration phenomena. Since stress raisers can reduce markedly the overall mechanical strength of structural components, engineers need reliable design methodologies suitable for performing the static assessment of notched 3D-printed objects accurately.

As far as static assessment of notched components is concerned, examination of the state of the art strongly supports the idea that the Theory of Critical Distances (TCD) [1] is the most powerful candidate to be used to design AM components against static loading since:

- the TCD assesses the detrimental effect of notches independently from their profile and sharpness;
- the TCD models the material morphology explicitly via suitable length scale parameters;
- the TCD can be used by directly postprocessing the results from simple linear-elastic Finite Element (FE) models (where the same numerical solid models can be used also to inform the manufacturing process).

In this scenario, the aim of the present paper is to review the research work we have supervised in recent years [2–7] to investigate whether the linear-elastic TCD is successful in performing the static assessment of notched 3D-printed components of PLA and ABS subjected to in service static loading.

2 Fundamentals of the Theory of Critical Distances (TCD)

The TCD postulates that static breakage of notched/cracked components subjected to Mode I loading takes place as soon as a critical distance-dependent effective stress, $\sigma_{\mathrm{eff}}(L)$, becomes equal to the material inherent strength, σ_0 [1, 8–11], i.e.:

$$\sigma_{\mathrm{eff}}(L) = \sigma_0 \Leftrightarrow \text{failure} \tag{1}$$

According to Eq. (1), the TCD is bi-parametrical design method where the critical distance, L, and the inherent material strength, σ_0, are the two material parameters being used to assess the extent of damage. In this setting, L is a length that is closely related to the micro/meso/macro-structural features of the material being designed.

An important feature of the TCD is that it performs static assessment by directly postprocessing the linear-elastic stress fields acting on the material in the vicinity of the notch being designed, with this holding true independently of the ductility level of the material under investigation [9, 10].

Under static loading, the TCD critical distance can directly be estimated via the following well-known relationship [1]:

$$L = \frac{1}{\pi}\left(\frac{K_{Ic}}{\sigma_0}\right)^2 \tag{2}$$

where K_{Ic} is the plane strain fracture toughness. Given the specific critical distance for the material under investigation, the design effective stress, $\sigma_{\text{eff}}(L)$, can be calculated directly according to either the Point Method, the Line Method or the Area Method as follows, respectively [1]:

$$\sigma_{\text{eff}}(L) = \sigma_y\left(\theta = 0, r = \frac{L}{2}\right) \quad \text{Point Method (Fig. 1b)} \tag{3}$$

$$\sigma_{\text{eff}}(L) = \frac{1}{2L}\int_0^{2L} \sigma_y(\theta = 0, r)\cdot \text{d}r \quad \text{Line Method (Fig. 1c)} \tag{4}$$

$$\sigma_{\text{eff}}(L) = \frac{4}{\pi L^2}\int_0^{\pi/2}\int_0^{L} \sigma_1(\theta, r)\cdot r\cdot \text{d}r\cdot \text{d}\theta \quad \text{Area Method (Fig. 1d)} \tag{5}$$

The meaning of the used symbols as well as the visual explanation of the effective stress determined according to definitions (3)–(5) are reported in Fig. 1a–d.

Equations (1) and (2) make it evident that inherent material strength σ_0 plays a role of primary importance when the TCD is employed to perform static assessment of notched components. In particular, while for brittle materials [1, 8], σ_0 is seen to approach the ultimate tensile strength, σ_{UTS}, for ductile materials (or nonlinear materials), σ_0 is instead quantified to be larger than σ_{UTS} [1, 9]. Further, σ_0 is seen to take on a value that is higher than σ_{UTS} also when the plain material fails by different mechanisms to those leading to the final breakage in the presence of notches [1]. These considerations make it evident that the most accurate way to determine σ_0 is by running specific experiments that involve notches whose presence results in different local stress distributions (Fig. 1e) [1, 9, 10].

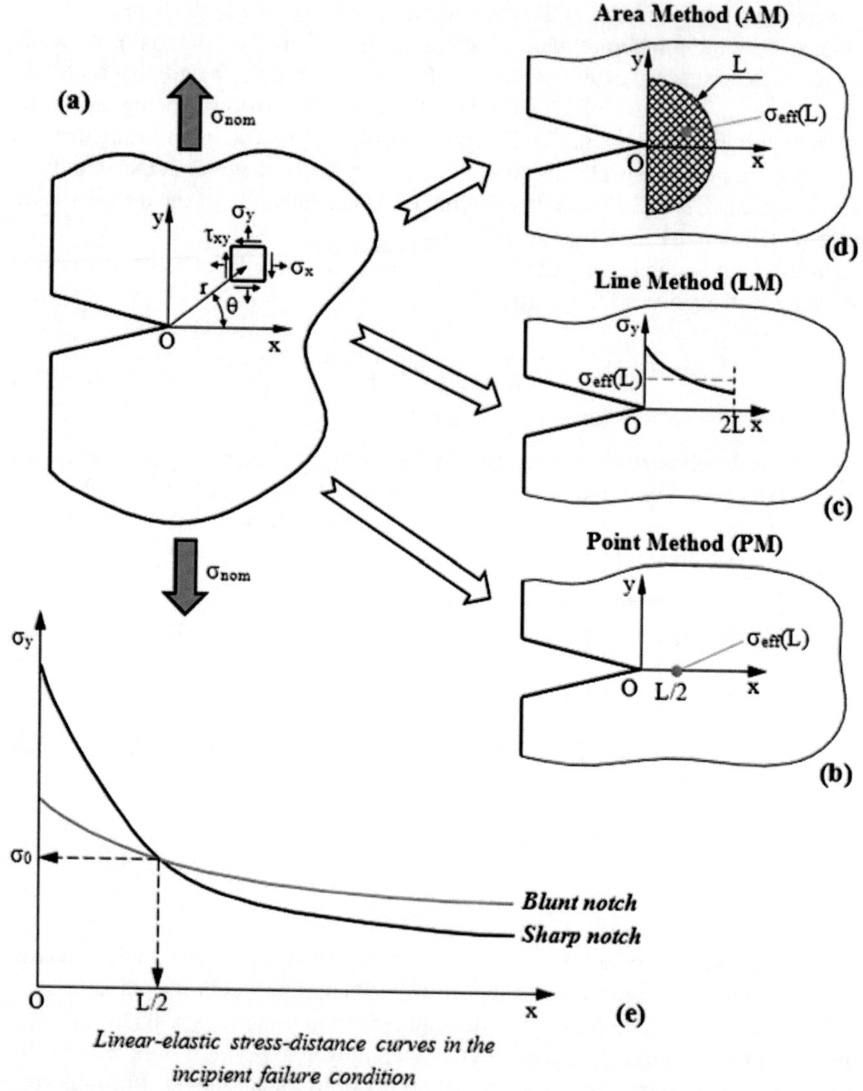

Fig. 1 Notch component and local system of coordinates (**a**); effective stress calculated according to the point method (**b**), to the line method (**c**) and to the area method (**d**); inherent strength σ_0 and critical distance L determined from experimental results generated by testing notches of different sharpness (**e**)

3 Fabrication and Testing of Laboratory Specimens

The specimens employed to assess the accuracy of the TCD in performing static assessment of notched AM polymers were fabricated by using 3D-printer Ultimaker 2 Extended+. White New Verbatim filaments and grey 750 g PRIMA filaments with initial diameter of 2.85 mm were employed to fabricate the samples of PLA [2] and ABS [5], respectively. For both materials, the specimens were manufactured flat on the build-plate using a 0.4 mm-diameter extrusion nozzle. For the PLA specimens, the nozzle temperature was set equal to 240 °C and the build-plate temperature to 60 °C [2]. In contrast, the ABS specimens were fabricated by setting the temperature of the nozzle equal to 255 °C and the temperature of the build-plate to 90 °C [5]. All the specimens were manufactured at a printing rate of 30 mm/s. The level of density was set equal to 100%, the height of the layers equal to 0.1 mm, and the thickness of the shell equal to 0.4 mm [2, 5]. As per Fig. 2, the plain and notched specimens were fabricated horizontally on the build-plate by making printing angle θ_p vary in the range 0°–90°. Since the used 3D-printer extruded the filaments forming the filling volume always at ± 45° to the principal manufacturing direction, setting angle θ_p equal either to 0° or to 90° returned specimens having a ± 45° lay-up. Using a similar stratagem, specimens with a −15°/75° lay-up were then fabricated by taking θ_p equal to both 30° and 60°, whereas samples with a 0°/90° lay-up were manufactured by taking θ_p equal to 45°. The technical drawings of the samples being fabricated and tested are seen in Fig. 3.

The un-notched samples and the specimens containing two opposite notches were tested under axial loading, whereas the samples with a single notch were tested under

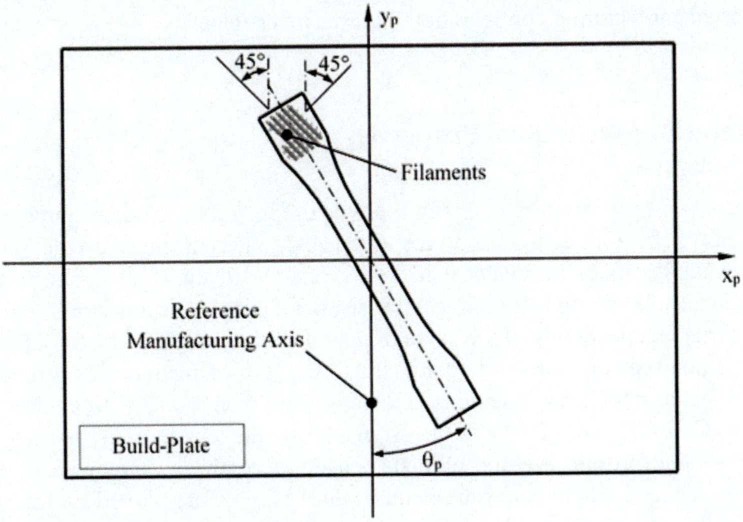

Fig. 2 Orientation of the deposition filaments

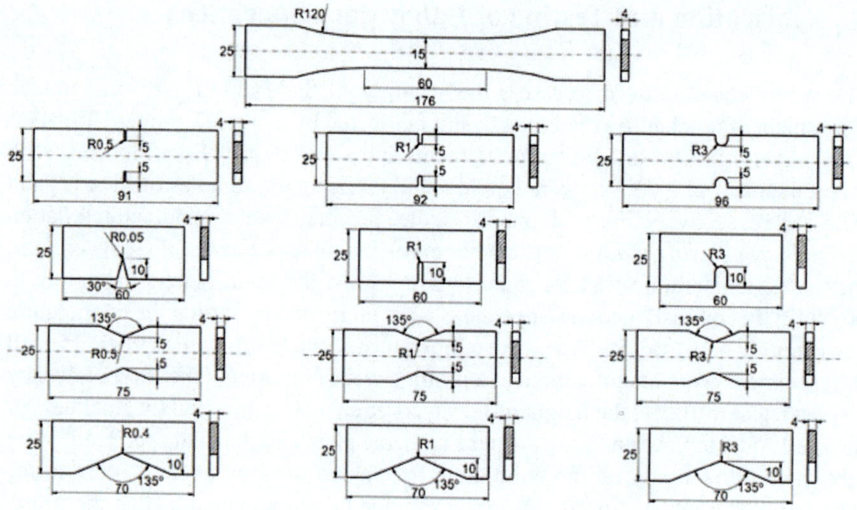

Fig. 3 Geometries and dimensions (in millimetres) of the tested specimens [2, 5]

three-point bending. For the bending tests, the span between the two lower supports was set equal to 50 mm for the specimens with U-notches and to 60 mm for the samples containing open notches.

All the specimens were tested under a displacement rate equal to 2 mm/min up to complete breakage by using a Shimadzu axial machine. Local axial strains in the un-notched samples were gathered at a frequency of 10 Hz via an extensometer having gauge length equal to 50 mm. Three different specimens were tested for any geometry/manufacturing configuration that was investigated.

4 Material Mechanical Properties

The stress, σ, versus strain, ε, diagrams reported in Fig. 4 show some examples of the mechanical behaviour displayed by the tested AM materials under tensile loading. These examples make it evident that the σ versus ε relationship was seen to be predominantly linear up to the maximum stress recorded during testing, with this holding true independently of the value of manufacturing angle θ_p [2, 5]. Taking as a starting point, this important outcome, the force–displacement curves determined from the plain specimens under tensile loading were then re-analysed to derive the elastic modulus, E, the 0.2% proof stress, $\sigma_{0.2\%}$, and the tensile strength, σ_{UTS}.

The obtained results for the ultimate tensile strength are summarised in the diagrams of Fig. 5 where the experimental values of σ_{UTS} are plotted against in-fill angle θ_p. These two diagrams make it evident that the σ_{UTS} values experimentally

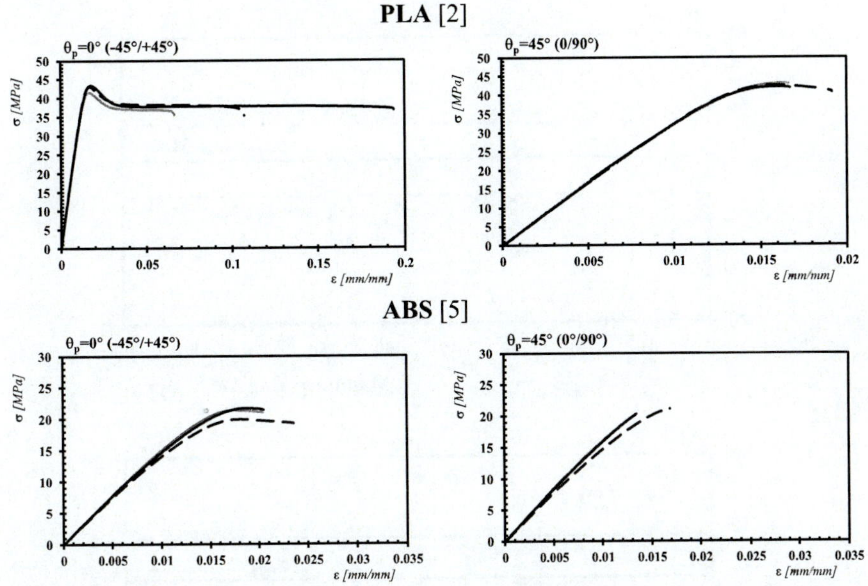

Fig. 4 Examples of the stress versus strain curves displayed by the tested materials for two different values of manufacturing angle θ_p [2, 5]

determined by making θ_p vary in the range 0°–90° were all within two standard deviations of the mean ($\pm$ 2S_D). Since both the 0.2% proof stress and Young's modulus were characterised by a similar trend, it was possible to conclude that the effect of θ_p on the mechanical behaviour of the two 3D-printed materials under investigation could be neglected with this resulting just in a little loss of accuracy [2, 5].

5 Overall Accuracy of the TCD

To use the TCD effective stress to assess the static strength of the notched samples shown in Fig. 2, the local linear-elastic stress fields in the notch regions were estimated by using commercial FE software ANSYS®. This was done by solving simple bi-dimensional linear-elastic FE modes, with the mesh density in the highly stressed regions being increased progressively until convergence occurred.

As far as ABS is concerned, a TCD critical distance, *L*, equal to 4.1 mm was estimated according to definition (2) by taking $\sigma_0 = \sigma_{UTS} = 23$ MPa and $K_{Ic} =$ 2.6 MPa·m$^{1/2}$ [5]. Turning to PLA, since it was not possible to determine a consistent value for the plane strain fracture toughness [2], *L* was directly estimated according to a procedure similar to the one shown in Fig. 1e. In particular, *L* was determined by averaging the distance resulting from the points of intersection between the straight horizontal lines modelling the plain material ultimate tensile strength ($\sigma_0 = \sigma_{UTS}$ in

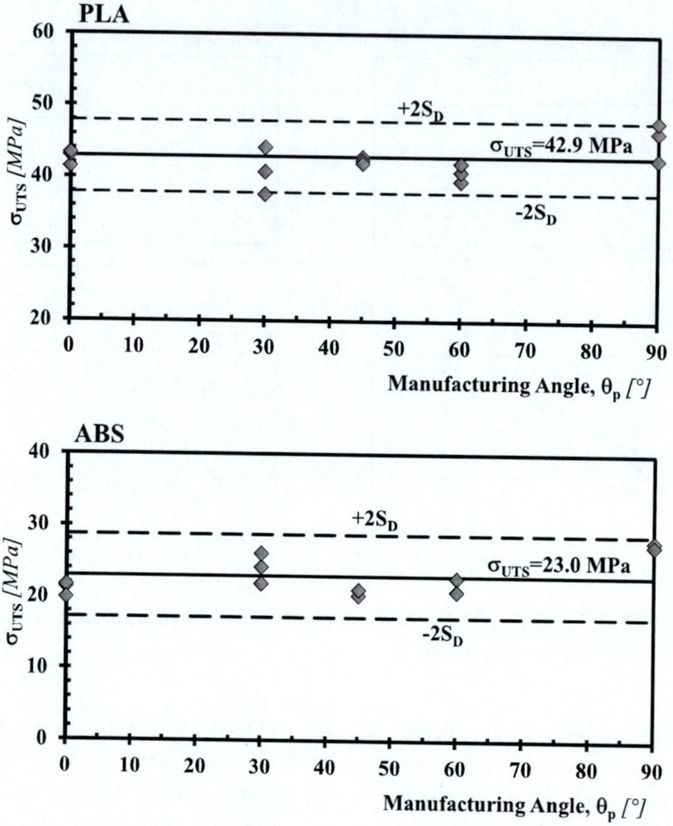

Fig. 5 Influence of manufacturing angle θ_p on the ultimate tensile strength of the two AM materials being investigated

Fig. 1e) and the stress-distance curves describing the local stress fields in the notched specimens (tested under tensile loading) having root radius, r_n, equal to 0.5 mm. This simple procedure resulted in an average value of the critical distance equal to 4.6 mm.

These two values for L allowed us to postprocess the notch results being generated according to the Point Method, Eq. (3), and the Area Method, Eq. (5). The Line Method, Eq. (4), instead could not be used because the integration length (i.e. $2L$) was larger than half net-width of the tested notched specimens [1].

The diagrams reported in Figs. 6 and 7 summarise the level of accuracy that was reached by using the TCD in terms of the Point and Area Method. In these charts, the error was calculated as:

$$\text{Error} = \frac{\sigma_{\text{eff}} - \sigma_{\text{UTS}}}{\sigma_{\text{UTS}}}\ [\%] \tag{6}$$

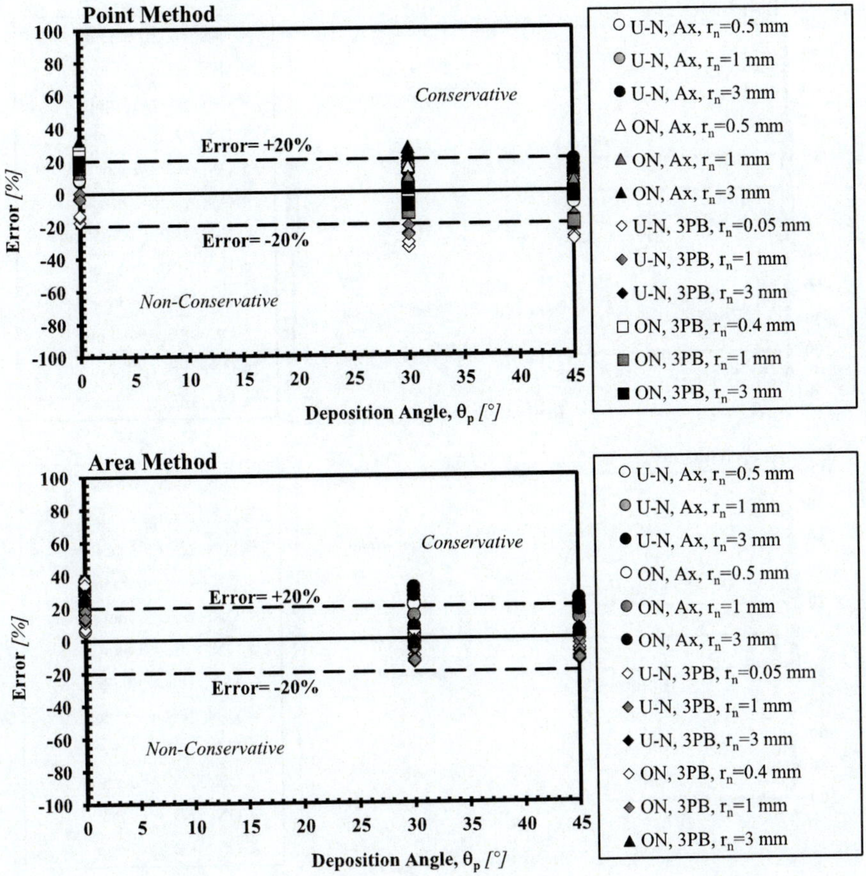

Fig. 6 Accuracy of the TCD used in the form of the Point and Area Method in estimating static strength of notched AM PLA (U-N = U-notched; ON = open notch; Ax = axial Loading; 3 PB = three-point bending; r_n = notch root radius) [2]

According to this definition, a positive value for the error denotes conservative estimates, whereas obviously, non-conservative predictions return negative errors.

The error diagrams reported Figs. 6 and 7 confirm that, as for other conventional engineering materials [1], the systematic usage of the Point and Line Method returned estimates mainly falling with in an error interval of ± 20%.

6 Conclusions

The present paper reviews the accuracy of the TCD in designing notched AM polymers against static loading. To consistently extend the use of this powerful theory to

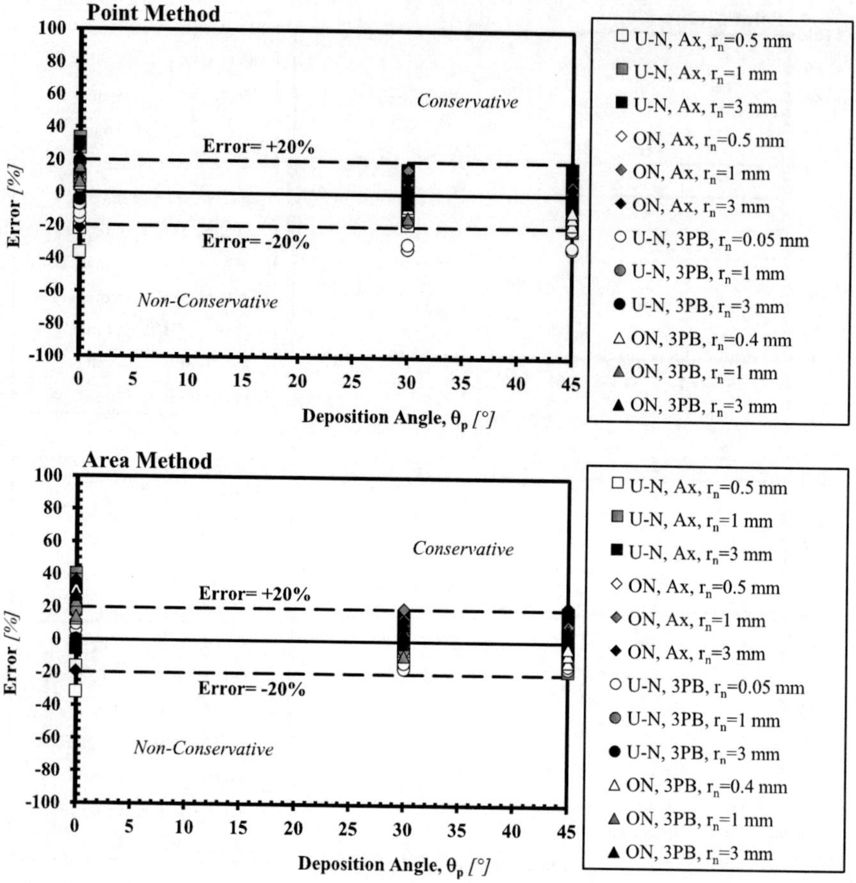

Fig. 7 Accuracy of the TCD used in the form of the Point and Area Method in estimating static strength of notched AM ABS (U-N = U-notched; ON = open notch; Ax = axial Loading; 3 PB = three-point bending; r_n = notch root radius) [5]

the static assessment of 3D-printed polymers wakened by stress raisers of different sharpness, the mechanical/cracking behaviour of two specific AM materials (i.e. PLA and ABS) were investigated by considering the effect of the manufacturing angle. The accuracy and reliability of the TCD was checked against a large number of experimental results generated by testing, under both tensile and bending loading, specimens of AM PLA as well as of AM ABS containing different geometrical features. Based on the work reviewed in the present paper, the most relevant conclusions are summarised in what follows.

- The linear-elastic TCD is successful in estimating static strength of notched AM PLA as well as of notched AM ABS.
- The TCD allows notched components of AM polymers to be designed against static loading by directly postprocessing the relevant stress fields determined

through conventional linear-elastic FE models. This implies that static assessment can be performed without the need for explicitly modelling the nonlinear mechanical behaviour displayed by AM polymers.

- With AM polymers as well, the TCD can be used to design notched components against static loading by treating the required critical distance as a material property whose value is not affected by the sharpness of the notch being assessed.

References

1. Taylor D (2007) The theory of critical distances: a new perspective in fracture mechanics. Elsevier Ltd., Oxford, UK
2. Ahmed AA, Susmel L (2018) A material length scale based methodology to assess static strength of notched additively manufactured polylactide (PLA). Fatigue Fract Eng Mater Struct 41(10):2071–2098
3. Ahmed AA, Susmel L (2019) Static assessment of plain/notched polylactide (PLA) 3D-printed with different in-fill levels: equivalent homogenised material concept and theory of critical distances. Fatigue Fract Eng Mater Struct 42:883–904
4. Ezeh OH, Susmel L (2019) Reference strength values to design against static and fatigue loading polylactide additively manufactured with in-fill level equal to 100%. Mat Des Process Commun e45:1–6
5. Ng CT, Susmel L (2020) Notch static strength of additively manufactured acrylonitrile butadiene styrene (ABS). Addit Manuf 34:101212
6. Seibert P, Susmel L, Berto F, Kästner M, Razavi SMJ (2021) Applicability of strain energy density criterion for fracture prediction of notched PLA specimens produced via fused deposition modelling. Eng Fract Mech 258:108103
7. Tu R, Gitman I, Susmel L (2022) Fuzzy inference system for failure strength estimation of plain and notched 3D-printed polylactide components. Fatigue Fract Eng Mater Struct 45:1663–1677
8. Susmel L, Taylor D (2008) The theory of critical distances to predict static strength of notched brittle components subjected to mixed-mode loading. Eng Frac Mech 75(3–4):534–550
9. Susmel L, Taylor D (2008) On the use of the theory of critical distances to predict static failures in ductile metallic materials containing different geometrical features. Eng Frac Mech 75:4410–4421
10. Susmel L, Taylor D (2010) The theory of critical Distances to estimate the static strength of notched samples of Al6082 loaded in combined tension and torsion. Part I: material cracking behaviour. Eng Frac Mech 77:452–469
11. Susmel L, Taylor D (2010) The theory of critical distances to estimate the static strength of notched samples of Al6082 loaded in combined tension and torsion. Part II: multiaxial static assessment. Eng Frac Mech 77:470–478

Fatigue

Modeling Crack Initiation in Low Cycle Fatigue: A Review

Hussein Zahran, Aleksandr Zinovev, Dmitry Terentyev, and Magd Abdel Wahab

Abstract The typical fatigue life of a component is mainly divided into two phases: crack initiation and crack propagation. This study is concerned with the crack initiation life as some designers regard the crack initiation as the end of the design life of the component. Crack initiation is caused by the formation of persistent slip bands that interact with the matrix leading to embryonic crack formation. There are several studies defining the point of crack initiation, and this article addresses some of these definitions. The main aim of this study is to review the different modeling methodologies for crack initiation under low cycle fatigue. These models are divided into three main types: microscopic models, damage parameters, and probabilistic models. There is no preferred methodology among the ones discussed. The choice of which model to use depends on the type of loading, the material in use, and the required level of detail. This study is intended as a reference for using one of these models or introducing modifications to enhance them.

Keywords Low cycle fatigue · Crack initiation · Damage parameters · Strain energy · SWT

1 Introduction

The study of fatigue loading for structural components is essential for several applications. In fact, fatigue damage is regarded as the most common damage mode for these components. The applications include aerospace, production machinery, and

H. Zahran (✉) · A. Zinovev · D. Terentyev
Structural Materials Group, Institute of Nuclear Materials Science, SCK CEN, 2400 Mol, Belgium
e-mail: hussein.zahran@sckcen.be

H. Zahran · M. Abdel Wahab
Soete Laboratory, Department of Electrical Energy, Faculty of Engineering and Architecture, Metals, Mechanical Constructions and Systems, Ghent University, 9000 Ghent, Belgium

H. Zahran
Department of Design and Production Engineering, Faculty of Engineering, Ain Shams University, Cairo 11517, Egypt

M. Abdel Wahab (ed.), *Proceedings of the 10th International Conference on Fracture Fatigue and Wear*, Lecture Notes in Mechanical Engineering,
https://doi.org/10.1007/978-981-19-7808-1_8

energy producing power plants. Fatigue loading is divided into two main types: Low cycle fatigue (LCF) and High cycle fatigue (HCF). This study is focused more on LCF loading, but the principles can be applied to both types.

Under LCF loading, the typical fatigue life of a component is divided into three phases [1, 2]:

(1) Crack initiation in which micro-cracks appear on the specimen surface. However, these cracks are retarded and none of them is dominant. The most dominant phases is lasting for 40–90% of the total fatigue life.
(2) Stable crack growth in which damage is localized by one dominant crack starting to propagate.
(3) Unstable crack growth, which is a rapid phase leading to the failure of the component.

This process is demonstrated in Fig. 1. It is worth noting that sometimes crack initiation stops at barriers such as grain boundaries and then the crack remains in this level and never reaches the critical size leading to the second or third phases.

Fatigue crack initiation takes place when a micro-crack propagates to an engineering measurement dimension [4]. For example, Bhattacharya and Ellingwood [5] defined a crack length of 0.076 mm as critical crack initiation size. Murtaza and Akid [6] defined it as 0.12 mm. Rodopoulos and De Los Rios [7] made a generalized assumption that crack initiation started when the crack length became several times the size of the microstructure.

Under this framework, the fatigue life of a smooth component can be calculated as follows:

$$N_f = N_i + N_p \tag{1}$$

Some studies provided a relation between the fatigue crack initiation point and the cyclic softening behavior of the material. These studies have been applied whenever the macroscopic detection of crack initiation is not possible. Zhang et al. [8] showed that crack initiation occurred at approximately 80% of the fatigue life under uniaxial loading based on damage mechanics theory. Another approach regards that crack initiation takes place at 10% drop in the stress amplitude [9, 10].

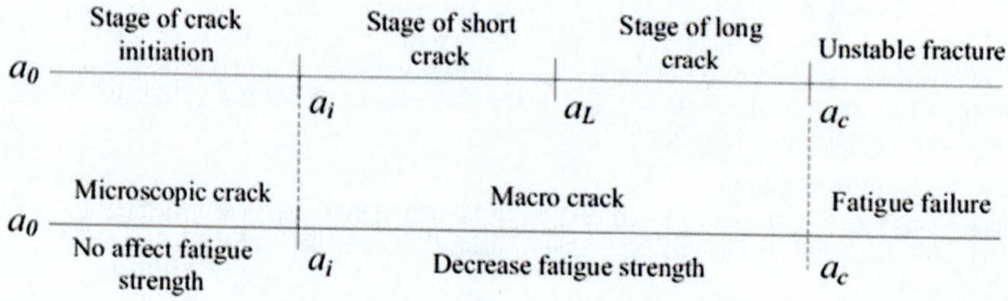

Fig. 1 Schematic diagram showing the fatigue failure process [3]

In this study, the mechanism of crack initiation is explained and then linked to the different macroscopic and microscopic modeling techniques.

2 Mechanism of Crack Initiation

Fatigue crack initiation is a random and complex process, which depends on the plastic strain amplitude, and the environment. The detailed process is summarized by Leon-Czares [11] as follows:

(1) Redundant dislocation density is generated due to fatigue hardening or softening to form a stabilized dislocation population.
(2) Constrained dislocation substructure is formed leading to the localization of slip into "Persistent Slip Bands (PSB)".
(3) The dislocation substructure interacts with a free surface producing extrusions and associated intrusions on it. This gives rise to the precipitation of free zones near the surface.
(4) Stress incompatibility is caused by this intrusion leading in the end to the production of embryonic cracks.

This process can be confirmed by microstructural investigations. Figure 2 shows the slip band formation, and the initiation of the crack at the interface between the slip band and matrix.

3 Crack Initiation Modeling

3.1 Microscopic Modeling

In order to represent the microstructure of the material, several techniques could be used. The immediate mapping of a microscopic image in a geometric finite element mesh is one of the most straightforward techniques, which leads to the matching between simulation and experimental results [13, 14]. The disadvantage of this method is that the simulation is strictly related to the mapped area. This is why other techniques have been developed to extend the simulation to the whole specimen. One technique is the statistical based synthetic microstructure using statistical information such as phase fraction and grain size distribution obtained from microscopic images. This information can be processed into a representative volume elements (RVE) model [15–18]. In the end, the mean material properties are given for the continuum.

A study by Tanaka and Mura [19] was one of the pioneering studies to present a model relating the fatigue crack initiation with the accumulation of dislocation dipoles that are generated by irreversible slip bands in one grain. In this model, the

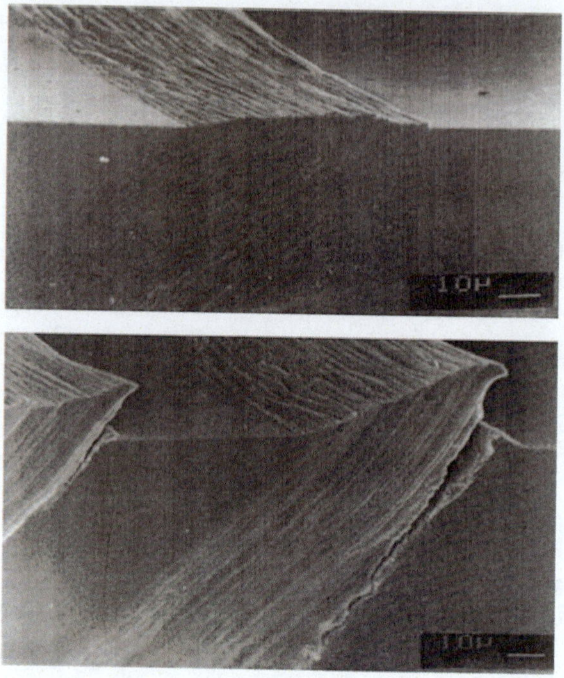

Fig. 2 Slip band formation as a result of fatigue loading followed by the formation of crack [12]

strain energy of dislocations is accumulated by the same amount during forward or reverse loading. A crack initiates when the accumulated energy reached fracture energy. The equation used to estimate crack initiation life is given by:

$$N_c = \frac{8GW_s}{\pi(1-\nu)d(\Delta\tau - 2k)^2} \tag{2}$$

where G is the shear modulus, ν is the Poisson's ratio, W_s is the specific fracture energy per unit area, and $\Delta\tau$ is the resolved shear stress range on the slip band. Also, k is the frictional stress of dislocation on the primary slip plane, and d is the length of the slip band. This model was used by Olfe et al. [20], and by Huang et al. [21] for strain controlled fatigue loading on ferrous and non-ferrous metals. It was also used by Hoshide and Kusuura [22] for multiaxial fatigue loading.

Crystal plasticity (CP) models have also been used in several studies to calculate the dislocation reactions on slip systems. It was initially developed for single crystals [23], but was then further enhanced to include polycrystalline features [24–27]. These features include dislocation density [28], grain boundary mechanisms [29–32], and damage initiation [33].

Gu et al. [34] used the crystal plasticity model to predict the crack initiation around the inclusions. In this model, the local accumulated dislocation p_{acc} was calculated using the following equation:

$$p_{acc} = \int_0^t \sqrt{\frac{2}{3} L_p : L_p \mathrm{d}t} \tag{3}$$

where L_p is the plastic velocity gradient that is calculated based on the plastic deformation gradient. This represents the shear stress, the slip systems, and the dislocation densities. This model allowed for the detection of the different crack nucleation sites in the microstructure.

3.2 Damage Parameters

Thermodynamic damage parameter. A study by Zhang et al. [35] was able to successfully simulate the fatigue crack initiation at the inclusion-matrix interface. This work was based on the thermodynamic theory of damage mechanics, in which the fatigue damage is regarded as an irreversible thermodynamic process [36]. This process dissipates the internal energy in the form of heat. A damage characterization parameter Υ is introduced, which is related to the material behavior as:

$$\Upsilon_{\max} = \frac{E}{2} \varepsilon_{\max}^2 \tag{4}$$

The damage characterization parameter can then be related to the time dependent damage D, representing the free energy dissipation caused by the internal damage, by the following equation:

$$\frac{\mathrm{d}D}{\mathrm{d}N} = aK\Upsilon_{\max}^{m/2} \tag{5}$$

The FE simulation of this study was performed on ANSYS using the Debonding submodule in the Contact Analysis module. This module is commonly used for cracking and separation of the material.

Strain energy damage parameter. The energy-based damage parameters usually possess specific physical significance, and more importantly, they can be used on high-cycle and low-cycle fatigue because they involve both stress and strain components [37].

A damage parameter was used by Yuenyong and Uthaisangsuk [38] merged with the Chaboche visco-plasticity model to detect the fatigue crack initiation. This damage parameter is based on the hysteresis energy criterion accounting for the accumulation of stabilized hysteresis strain energy. This damage criterion was used in several other studies [39–41]. The accumulated inelastic strain energy per cycle ΔW with the number of cycles N is formulated by the following equation:

$$N = A\Delta W^B \tag{6}$$

In which A and B are material parameters. The progression of the damage parameter with the number of cycles is given by:

$$\frac{dD}{dN} = \frac{1}{A\Delta W^B} \tag{7}$$

This damage parameter affects the stress tensor (σ) causing the material softening progressively damaged until failure when the value of D reaches 1. The effect is given as:

$$\sigma = (1 - D)\overline{\sigma} \tag{8}$$

Critical plane (CP) approach. Strain energy is scalar quantity, and this is why most energy-based models cannot solely describe crack initiation and propagation on a specific plane for most metallic materials [42, 43]. This raises the need for the critical plane (CP) approach to detect the favorable plane for the crack to initiate. Smith et al. [44] proposed a damage parameter to describe the crack initiation in a specific plane. This parameter depends on the maximum normal stress, and the true normal strain range on a specific plane as given by:

$$P_{\text{SWT}} = \sigma_{t,\max}\frac{\Delta\varepsilon_{t,T}}{2} \tag{9}$$

where $\sigma_{t,\max}$ and $\Delta\varepsilon_t$ are the maximum normal stress and the true normal strain range, respectively.

The SWT parameter was proved to be incapable of detecting damage in many materials, especially the ones dominated by shear crack initiation [45, 46].

Another damage parameter was proposed by Findley [47] taking into account the effect of the normal stress on the maximum shear stress plane. The introduced critical plane model predicts the fatigue crack plane with orientation θ having maximum Findley Damage parameter. This is expressed by:

$$f(\theta) = \frac{\Delta\tau}{2} + k\sigma_n \tag{10}$$

where $\Delta\tau/2$ is the shear stress amplitude on a plane with orientation θ, σ_n is the maximum normal stress on that plane, and k is a material parameter.

Several versions of this damage parameter have been developed. Brown and Miller considered both the shear and normal strain on the plane of maximum shear stress. This is based on the assumption that cyclic shear strain will help in the crack nucleation, and the normal strain will help in its growth. A simplified formula was proposed by Kandil et al. [48] given by:

$$\frac{\Delta\hat{\gamma}}{2} = \frac{\Delta\gamma_{\max}}{2} + S\Delta\varepsilon_n \tag{11}$$

where $\Delta\gamma_{\max}$ is the maximum shear strain range, and $\Delta\varepsilon_n$ is the maximum normal strain range on the plane with the maximum shear strain range. S is a material parameter representing the influence of the normal strain on crack growth.

The different critical plane methods have the benefit of being used to model multiaxial fatigue loading. This is why these methods have been used in various studies [49, 50].

3.3 Probabilistic Modeling

As discussed previously, crack initiation process is a random process that depends on several uncertainties such as geometrical features, and microstructures (presence of inclusions or defects). This is why sometimes deterministic models become inaccurate especially on the attempt of generalizing these models.

Some studies have been calling for some statistical arguments in order to apply the different models for design purposes. From here, probabilistic models were established in order to carry out reliability analysis. A study by Castillo and Fernandez-Canteli proposed a probabilistic model based on Weibull or Gumbel distributions [51]. This model starts with one damage parameter such as the life to initiate a macrocrack N_f and expressed by:

$$\psi = q(N_f) \tag{12}$$

ψ here represents the fatigue damage parameter, and q represents a decreasing function of total life in terms of reversals to macro-crack initiation. This damage parameter is then coupled with a probabilistic Weibull regression model in the following compact form:

$$f(\psi)f(N_f) = \nu \tag{13}$$

where ν here is a regression parameter and f is logarithmic function with the following form:

$$f(x) = \log(x) - \theta_i = \log\left(\frac{x}{e^{\theta_i}}\right) \tag{14}$$

In this equation, $i = 1, 2, \theta_1 = \psi_0$, and $\theta_2 = N_0$. These parameters can be extracted from experimental data. Then, a set of values ν_k (where $k = 1, 2, \ldots$ till the number of experiments) are fitted with a three-parameter Weibull distribution (Fig. 3):

$$\mathrm{N}_k \sim W(\lambda, \delta, \beta) \tag{15}$$

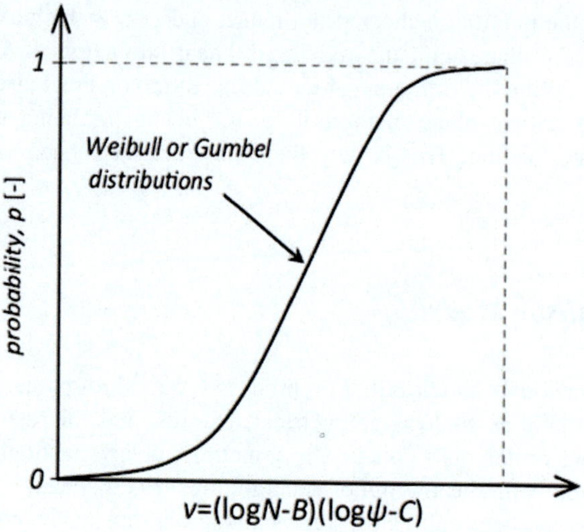

Fig. 3 Schematic diagram of the probability of regression parameter [52]

The parameters of this distribution are: λ defining the position of the corresponding zero percentile hyperbola, δ defining the scale factor, and β representing the Weibull shape parameter of the cumulative distribution variable.

The advantage of this probabilistic model is that it can be coupled with various damage parameter models turning it from deterministic to probabilistic. It has successfully been used in combination with Chaboche deterministic method [53], multiaxial fatigue virtual strain energy model [54], and Miner's linear damage summation [55]. The model was also used to estimate fatigue life on structural components.

Other probabilistic approaches have been studied such as the model by Zhu et al. [56] using Bayes theorem, and Koutiri et al. [57] using the weakest link concept.

4 Conclusion

In this review, several modeling methodologies have been discussed to predict the low cycle fatigue crack initiation lifetime. These methodologies can be microscopic, macroscopic, or statistics based. Some models were able to cover different applications: low cycle fatigue, high cycle fatigue, uniaxial loading, multiaxial loading, etc., while other models had limitations. Also, not every model can be applied on every type of material, so care must be made when choosing the suitable model. In the end, there is no preferred model out of the ones in discussion. It is also important to mention that there is still room for improvement in these models and correlating them with the physics of crack initiation. As discussed, crack initiation is a complex and random process with no agreed upon definition.

Acknowledgements This work has been carried out within the framework of the EUROfusion Consortium, funded by the European Union via the Euratom Research and Training Programme (Grant Agreement No 101052200—EUROfusion). Views and opinions expressed are however those of the author(s) only and do not necessarily reflect those of the European Union or the European Commission. Neither the European Union nor the European Commission can be held responsible for them.

References

1. Makkonen M (2009) Predicting the total fatigue life in metals. Int J Fatigue 31(7):1163–1175
2. Jiang Z, Han Z, Li M (2022) A probabilistic model for low-cycle fatigue crack initiation under variable load cycles. Int J Fatigue 155:106528
3. Wang H et al (2019) Numerical method for estimating fatigue crack initiation size using elastic–plastic fracture mechanics method. Appl Math Model 73:365–377
4. Radaj D (1996) Review of fatigue strength assessment of nonwelded and welded structures based on local parameters. Int J Fatigue 18(3):153–170
5. Bhattacharya B, Ellingwood B (1998) Continuum damage mechanics analysis of fatigue crack initiation. Int J Fatigue 20(9):631–639
6. Murtaza G, Akid R (1995) Modelling short fatigue crack growth in a heat-treated low-alloy steel. Int J Fatigue 17(3):207–214
7. Rodopoulos C, De Los Rios E (2002) Theoretical analysis on the behaviour of short fatigue cracks. Int J Fatigue 24(7):719–724
8. Zhang Y, Gao L, Wang C (2013) A prediction model for low cycle fatigue crack initiation under axial loading. In: Proceedings of 13th International Conference of Fracture (ICF 2013)
9. Mitchell MR (1992) Advances in fatigue lifetime predictive techniques, vol 1122. ASTM International
10. ASTM (2012) Standard test method for strain-controlled fatigue testing. ASTM Standard E606
11. León-Cázares F et al (2020) Stress response and microstructural evolution of nickel-based superalloys during low cycle fatigue: physics-based modelling of cyclic hardening and softening. Int J Plast 128:102682
12. Basinski Z, Basinski S (1992) Fundamental aspects of low amplitude cyclic deformation in face-centred cubic crystals. Prog Mater Sci 36:89–148
13. Lian J et al (2014) A method to quantitatively upscale the damage initiation of dual-phase steels under various stress states from microscale to macroscale. Comput Mater Sci 94:245–257
14. Özden UA et al (2015) Mesoscopical finite element simulation of fatigue crack propagation in WC/Co-hardmetal. Int J Refract Metal Hard Mater 49:261–267
15. Nygårds M, Gudmundson P (2002) Three-dimensional periodic Voronoi grain models and micromechanical FE-simulations of a two-phase steel. Comput Mater Sci 24(4):513–519
16. Nygårds M, Gudmundson P (2002) Micromechanical modeling of ferritic/pearlitic steels. Mater Sci Eng A 325(1–2):435–443
17. Zhao Y, Tryon R (2004) Automatic 3-D simulation and micro-stress distribution of polycrystalline metallic materials. Comput Methods Appl Mech Eng 193(36–38):3919–3934
18. Vajragupta N et al (2014) The modeling scheme to evaluate the influence of microstructure features on microcrack formation of DP-steel: the artificial microstructure model and its application to predict the strain hardening behavior. Comput Mater Sci 94:198–213
19. Tanaka K, Mura T (1981) A dislocation model for fatigue crack initiation
20. Olfe J, Zimmermann A, Rie K (2000) Simulation of microcrack formation and growth during cyclic loading considering microstructure
21. Huang X et al (2007) Simplified three-dimensional model for fatigue crack initiation. Eng Fract Mech 74(18):2981–2991

22. Hoshide T, Kusuura K (1998) Life prediction by simulation of crack growth in notched components with different microstructures and under multiaxial fatigue. Fatigue Fract Eng Mater Struct 21(2):201–213
23. Becker R (1991) The effects of shear constraints on the lattice rotation of FCC crystals in (011) channel-die compression. J Mech Phys Solids 39(4):459–476
24. Raabe D, Roters F (2004) Using texture components in crystal plasticity finite element simulations. Int J Plast 20(3):339–361
25. Zisu Z et al (2001) Introduction of a texture component crystal plasticity finite element method for anisotropy simulations. Adv Eng Mater 3
26. Melchior MA, Delannay L (2006) A texture discretization technique adapted to polycrystalline aggregates with non-uniform grain size. Comput Mater Sci 37(4):557–564
27. Eisenlohr P, Roters F (2008) Selecting a set of discrete orientations for accurate texture reconstruction. Comput Mater Sci 42(4):670–678
28. Arsenlis A et al (2004) On the evolution of crystallographic dislocation density in non-homogeneously deforming crystals. J Mech Phys Solids 52(6):1213–1246
29. Ma A, Roters F, Raabe D (2006) A dislocation density based constitutive model for crystal plasticity FEM including geometrically necessary dislocations. Acta Mater 54(8):2169–2179
30. Ma A, Roters F, Raabe D (2006) Studying the effect of grain boundaries in dislocation density based crystal-plasticity finite element simulations. Int J Solids Struct 43(24):7287–7303
31. Ma A, Roters F, Raabe D (2006) On the consideration of interactions between dislocations and grain boundaries in crystal plasticity finite element modelling—theory, experiments, and simulations. Acta Mater 54(8):2181–2194
32. Knezevic M et al (2014) Three dimensional predictions of grain scale plasticity and grain boundaries using crystal plasticity finite element models. Comput Methods Appl Mech Eng 277:239–259
33. Bieler T et al (2009) The role of heterogeneous deformation on damage nucleation at grain boundaries in single phase metals. Int J Plast 25(9):1655–1683
34. Gu C et al (2019) Microstructure-based fatigue modelling with residual stresses: prediction of the microcrack initiation around inclusions. Mater Sci Eng A 751:133–141
35. Zhang S et al (2021) Low-cycle fatigue crack initiation simulation and life prediction of powder superalloy considering inclusion-matrix interface debonding. Materials 14(14):4018
36. Lemaitre J, Desmorat R (2006) Engineering damage mechanics: ductile, creep, fatigue and brittle failures. Springer Science & Business Media
37. Zhu H et al (2019) A novel energy-based equivalent damage parameter for multiaxial fatigue life prediction. Int J Fatigue 121:1–8
38. Yuenyong J, Uthaisangsuk V (2020) Micromechanics based modelling of fatigue crack initiation of high strength steel. Int J Fatigue 139:105762
39. Darveaux R (2002) Effect of simulation methodology on solder joint crack growth correlation and fatigue life prediction. J Electron Packag 124(3):147–154
40. Lau JH, Pan SH, Chang C (2002) A new thermal-fatigue life prediction model for wafer level chip scale package (WLCSP) solder joints. J Electron Packag 124(3):212–220
41. Kliman V, Bílý M (1984) Hysteresis energy of cyclic loading. Mater Sci Eng 68(1):11–18
42. Wang C, Brown M (1993) A path-independent parameter for fatigue under proportional and non-proportional loading. Fatigue Fract Eng Mater Struct 16(12):1285–1297
43. Fatemi A, Shamsaei N (2011) Multiaxial fatigue: an overview and some approximation models for life estimation. Int J Fatigue 33(8):948–958
44. Smith KN, Watson P, Topper TH (1970) Stress-strain function for the fatigue of metals. J Mater 5(4):767–778
45. Kalnaus S, Jiang Y (2008) Fatigue of AL6XN stainless steel. J Eng Mater Technol 130(3)
46. Wu Z-R, Hu X-T, Song Y-D (2014) Multiaxial fatigue life prediction for titanium alloy TC4 under proportional and nonproportional loading. Int J Fatigue 59:170–175
47. Findley WN (1959) A theory for the effect of mean stress on fatigue of metals under combined torsion and axial load or bending. J Eng Ind 81(4):301–305

48. Kandil FA, Brown MW, Miller K (1982) Biaxial low-cycle fatigue failure of 316 stainless steel at elevated temperatures. In: Mechanical behaviour and nuclear applications of stainless steel at elevated temperatures
49. Gan L, Wu H, Zhong Z (2019) Use of an energy-based/critical plane model to assess fatigue life under low-cycle multiaxial cycles. Fatigue Fract Eng Mater Struct 42(12):2694–2708
50. Ono Y et al (2022) Damage-based assessment of the fatigue crack initiation site in high-strength steel welded joints treated by HFMI. Metals 12(1):145
51. Weibull W (1949) A statistical report of fatigue failure in solids. Transactions 27
52. Correia J et al (2017) Generalized probabilistic model allowing for various fatigue damage variables. Int J Fatigue 100:187–194
53. Amraoui A, Gharad A, Bensalah MO (2013) Probabilistic model of damage from a deterministic model and estimated lifetimes (case SAE 1137 carbon steel). Appl Math Sci 7(90):4465–4474
54. Calvo S et al (2011) Probabilistic formulation of the multiaxial fatigue damage of Liu. Int J Fatigue 33(3):460–465
55. Blasón S et al (2016) A probabilistic analysis of Miner's law for different loading conditions. Struct Eng Mech 60(1):71–90
56. Zhu S-P et al (2012) Probabilistic low cycle fatigue life prediction using an energy-based damage parameter and accounting for model uncertainty. Int J Damage Mech 21(8):1128–1153
57. Koutiri I et al (2013) A probabilistic model for the high cycle fatigue behaviour of cast aluminium alloys subject to complex loads. Int J Fatigue 47:137–147

Investigating the Effect of the Characteristics of Inclusions and Micro-voids on Fretting Fatigue Initiation Lifetime with Two-Scale Analysis Approach

Can Wang and Magd Abdel Wahab

Abstract Fretting fatigue problem involves a multiaxial stress state due to the contact problem. This complicated mechanical situation decreases the total fatigue life seriously. To obtain an accurate estimation of fretting fatigue lifetime, continuum damage mechanics (CDM) approach in conjunction with finite element method (FEM) is applied in this paper. Usually, the material is assumed by researchers as homogeneous material, but some micro-voids, inclusions and micro-cracks exist in the material in reality. These defects influence the estimation of fretting fatigue lifetime dramatically, and it is necessary to take these defects into account. In this paper, the authors investigate the characteristics of micro-voids and inclusions on fretting fatigue lifetime with a two-scale analysis approach.

Keywords Fretting fatigue · Continuum damage mechanics · Defects · Two-scale analysis

1 Introduction

Fretting fatigue damage occurs at the position of the contact body because of the oscillation force that causes small movement. In addition, fretting fatigue damage usually occurs on bolts and riveted connections or components supporting vibration movement. Under fretting conditions, the fatigue life will be shortened seriously [1, 2]. Fretting fatigue failure process is divided into two stages, namely, crack initiation and crack propagation. At the same time, considering the fatigue conditions, most of the total life was consumed in the nucleation part [3].

As mentioned earlier, the crack initiation stage is important. However, it is difficult to obtain the crack initiation point and crack initiation life from the experimental results. Therefore, finite element method (FEM) has been widely used in this field. It can effectively simulate contact conditions by reducing time and provide relatively accurate prediction results in combination with other prediction methods. Several

C. Wang (✉) · M. Abdel Wahab
Soete Laboratory, Department of Electrical Energy, Metals, Mechanical Constructions and Systems, Faculty of Engineering and Architecture, Ghent University, Ghent, Belgium
e-mail: can.wang@ugent.be

M. Abdel Wahab (ed.), *Proceedings of the 10th International Conference on Fracture Fatigue and Wear*, Lecture Notes in Mechanical Engineering,
https://doi.org/10.1007/978-981-19-7808-1_9

Table 1 The fretting fatigue experimental data

Test No.	σ_{axial} (MPa)	Q (N)	F (N)	$N_{failure}$ (cycle)
1	115	186.25	543	1,105,245

methods and damage parameters are proposed in the literature to give the crack initiation point, crack direction and crack initiation life [4].

Kachanov [5] first proposed continuous damage mechanics (CDM) based on thermodynamic principles. In this method, a damage variable D is introduced, and it can be combined with the effective stress to estimate the damage evolution. Some researchers have extended this method to various fields. Some researchers [6] applied it to high cycle fatigue conditions and achieved good results. CDM method [7, 8] has successfully solved the problems of low cycle fatigue and fatigue creep. CDM method was applied to solve fatigue and fretting fatigue problems by some researchers [9, 10].

The heterogeneity of material will dramatically influence the fretting fatigue lifetime estimation with FEM. The existence of micro-voids and inclusions will decrease the fretting fatigue lifetime in real experimental conditions, it is necessary to take the defects into account when we estimate the fretting fatigue lifetime with FEM. Thus, in this paper, the authors will consider the effect of the characteristics of inclusions and micro-voids on fretting fatigue initiation lifetime with two-scale analysis approach.

2 Materials and Experiments

The present study refers to the experiment which was conducted by Hojjati-Talemi and Abdel Wahab [11] and the details are listed in Table 1. The schematic of the experiment setup is shown in Fig. 1. As is shown, in the x-direction, the flat specimen sustains the axial cycle loading and friction force. In the y-direction, it sustains static normal force from the cylindrical pad.

The material of cylindrical pad and flat specimen is AA2024-T3. According to some studies, there are two kinds of inclusions (Al_2O_3 and Al_2CuMg) and micropores in the material. Material properties are taken from literature [12, 13], and the size of these defects ranges from 23 to 65 μm. This paper takes the aspect ratio of 1 as an example. Table 2 shows the material properties and volume fraction of these defects.

The stress ratios of axial stress and tangential stress are 0.1 and −1, respectively.

3 Continuum Damage Mechanics Method

Kachanov [5] proposed the CDM approach based on thermodynamic theory. Hojjati-Talemi and Abdel Wahab [11] introduced the damage parameter D in detail. The

Fig. 1 Schematic of fretting fatigue test setup

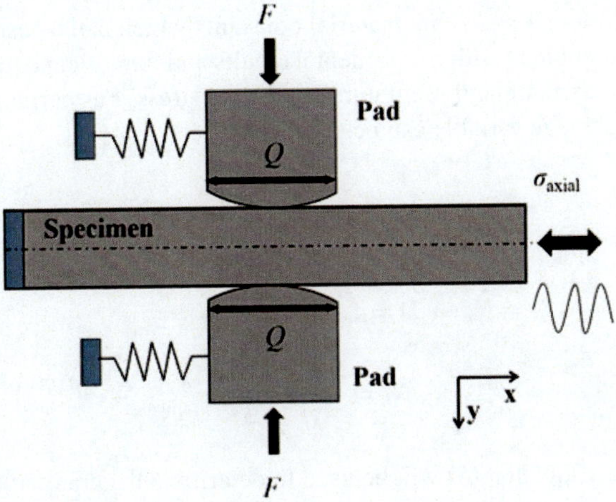

Table 2 The material properties and defects characteristics

Material	*E* (GPa)	*v* (−)	Volume fraction (%)	Size (μm)
AA 2024-T3	72.1	0.33		
Al_2O_3	380	0.2	1.35	23–65
Al_2CuMg	120.5	0.2	1.35	23–65
Micro-voids	–		0.7	23–65

parameter determining the crack initiation point is the equivalent multiaxial stress (σ *), and it can be estimated by Von Mises equivalent stress (σ_{eq}) and triaxiality function (R_v), see Eq. (1) [14].

$$\sigma^* = \sigma_{\text{eq}}(R_v)^{1/2} \tag{1}$$

where

$$R_v = \frac{2}{3}(1+v) + 3(1-2v)\left(\frac{\sigma_H}{\sigma_{\text{eq}}}\right)^2 \tag{2}$$

Herein, σ_H is the hydrostatic stress.

Hojjati-Talemi and Abdel Wahab [15] extended it to predict the crack initiation lifetime under fretting fatigue condition. The integration relationship between damage parameter D and number of cycles N is presented in Eq. (3).

$$\frac{\partial D}{\partial N} = A\frac{\left(\sigma_{\text{eq,max}}^{\beta+2} - \sigma_{\text{eq,min}}^{\beta+2}\right)}{(1-D)^{\beta+2}} R_v^{\left(\frac{\beta}{2}\right)+1} \tag{3}$$

where A and β are material constants which can be estimated by regression method combined with experimental results and numerical data [11], $\sigma_{\text{eq, max}}$, $\sigma_{\text{eq, min}}$ means maximum and minimum Von Mises stress. Furthermore, by integrating Eq. (3), the damage variable can be expressed as:

$$D = 1 - \left[1 - A(\beta + 3)\left(\sigma_{\text{eq,max}}^{\beta+2} - \sigma_{\text{eq,min}}^{\beta+2}\right) R_v^{\left(\frac{\beta}{2}\right)+1} N\right]^{\frac{1}{(\beta+3)}} \quad (4)$$

At $N = N_{\text{i}} \rightarrow D = 1$:

$$N_i = \frac{1}{A(\beta + 3)}\left(\sigma_{\text{eq,max}}^{\beta+2} - \sigma_{\text{eq,min}}^{\beta+2}\right)^{-1} R_v^{-\left(\frac{\beta}{2}\right)-1} \quad (5)$$

Equation (5) will be used to determine the crack initiation life.

4 Two-Scale Numerical Models

In two-scale numerical method, there are two kinds of models: macroscopic model and microscopic model. The macroscopic model is applied to solve the contact problem and obtain the boundary condition for the microscopic models, meanwhile, the authors defined the homogeneous material in macroscopic one. All these models are constructed in commercial software ABAQUS.

It can be seen from Fig. 2 that the y-direction of the specimen is constrained by the bottom, and the x-direction of the pad is constrained by the left and right sides. Axial cyclic stress and reaction cyclic stress are applied on the right side and the left side of the specimen, respectively. The value of reaction cyclic stress (σ_{reaction}) can be calculated with Eq. (6).

$$\sigma_{\text{reaction}} = \sigma_{\text{axial}} \frac{Q}{A_s} \quad (6)$$

where A_{s} means the cross-section area of specimen.

The small region A is the most damaged region in this model thus the boundary conditions from this region will be applied to microscopic models to investigate the effect of the characteristics of inclusions and micro-voids on fretting fatigue initiation lifetime with two-scale analysis approach. A technique in ABAQUS called sub-model technique which can heritage the boundary condition from the macroscopic model is applied in the work.

In this paper, the main point is to know what's the serious factors decrease the fretting fatigue lifetime, the sizes of micro-voids and inclusions, the volume fractions of micro-voids and inclusions, or the kinds of inclusions. Therefore, several microscopic models are constructed due to the research objective which are shown in Figs. 3 and 4.

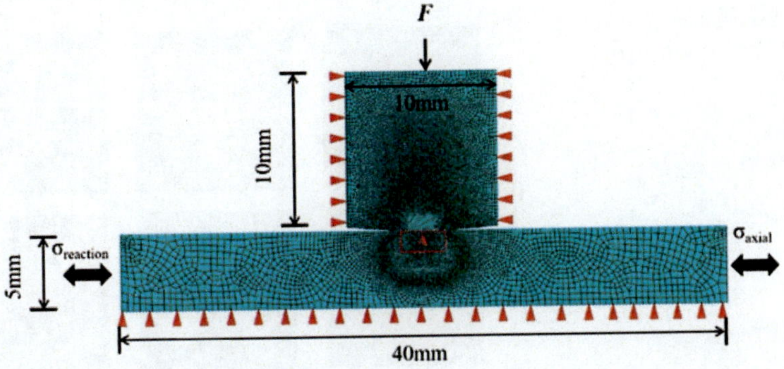

Fig. 2 Homogeneous macroscopic model

Fig. 3 Sketch map of microscopic models: **a** $r = 12.6\ \mu m$; **b** $r = 16.3\ \mu m$ and **c** $r = 28.2\ \mu m$

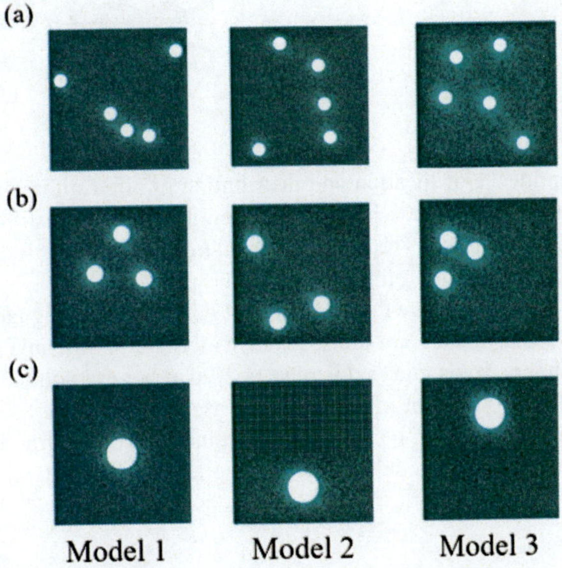

Figure 2 shows the macroscopic model, the width and length of the cylindrical pad are 10 mm, the length and the width of the specimen are 40 mm and 5 mm, the thickness of both parts is 4 mm. The size of region A is 1 mm × 0.25 mm, the authors divide this region into four microscopic models, each microscopic model's size is 250 μm × 250 μm. The element type is CPE4R.

Figure 3 shows the microscopic models which aim to investigate the effect of sizes of micro-voids and inclusions (the radius of defects r) in fretting fatigue lifetime. In Fig. 3, the authors just take the microscopic models of micro-voids as examples, and it needs to be clear, that the distributions and locations of inclusions are the same as the models in Fig. 3. In these models, the volume fraction of micro-voids and inclusions maintain 4%, the difference only occurs at the sizes. To avoid the

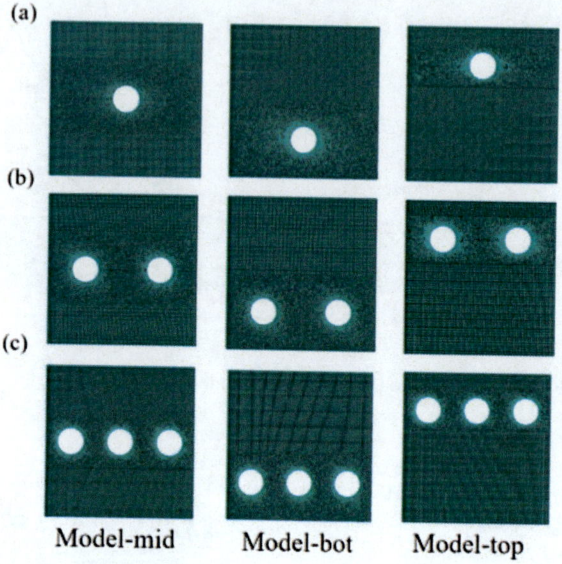

Fig. 4 Sketch map of microscopic models: **a** $vf = 2.04\%$ **b** $vf = 4.08\%$ and **c** $vf = 6.12\%$

influence of location and distribution on life estimation, three microscopic models are constructed, which are different locations or distributions of defects under each size condition. The mesh size in microscopic models is 2 μm and the elements nearby the defects are refined as 0.5 μm.

To investigate the size effect of defects on fretting fatigue lifetime, several microscopic models are constructed which are shown in Fig. 4. The radius of defects (micro-voids and inclusions) is 20 μm but the volume fractions (vf) are different. The positions of micro-voids are as same as that of two kinds of inclusions (Al_2O_3 and Al_2CuMg). It can be seen that in each volume fraction condition, there are three microscopic models which are applied to avoid the influence of defects locations.

5 Results and Discussion

5.1 *Distribution of Damage Parameter and Stress*

Since the damage parameter and stress peaks are located in the region $a/2$ to a, the stress and damage parameter contour plots in the fourth part of the micromodel which are shown here.

The peak values of damage parameters and Von Mises stress are all locate at the edge of the micro-voids (see Fig. 5). The existence of micro-voids affects the distributions of stress distribution, and the initiation point would possibly shift from the contact surface to the edge of micro-voids. Comparing with the results from

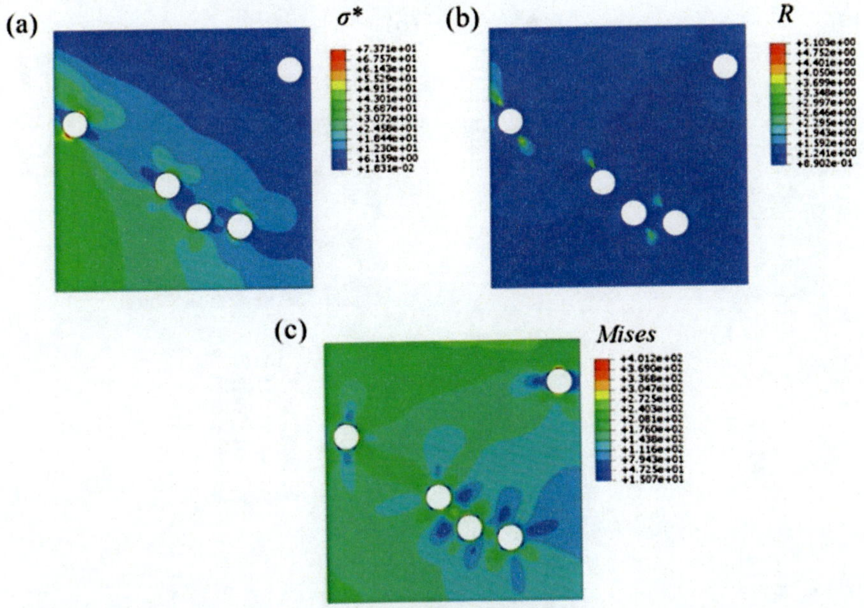

Fig. 5 Distribution of damage parameters and stress of models containing micro-voids: **a** equivalent multiaxial stress, **b** triaxiality function, and **c** Von Mises stress

Fig. 6, the maximum equivalent multiaxial stress (σ^*), triaxiality function (R) and the Von Mises stress are higher than that of models containing inclusions.

To get crack initiation lifetime, average method needs to be applied, in this part it will not be introduced in detail, the average method refers to the previous study [16]. The comparison of crack initiation lifetime in different microscopic models will be introduced in the following sections.

5.2 Effect of Defects Sizes on Fretting Fatigue Lifetime

It needs to be clarified that the $N_{i,p}$ means the predicted fretting fatigue initiation lifetime, and the $N_{i,e}$ is the experimental fretting fatigue initiation lifetime which is obtained by subtracting the propagation lifetime from the total lifetime. Meanwhile, the propagation lifetime refers to the literature [17].

Figure 7a shows the life ratio of models containing three different sizes of micro-voids. The distributions of micro-voids are different from each other in the three models (Model 1, Model 2 and Model 3) which are shown in Fig. 2. It can be seen that the lifetimes vary from the distributions of micro-voids of the same size. When the size of micro-voids is 12.6 μm, the range of life ratios ($N_{i,p}/N_{i,e}$) is from 0.45 to 0.8, the difference arrives at 0.35. When the size is 28.2 μm, life ratios fluctuate between 0.45 and 0.7. From the green triangles in Fig. 7a, the average life ratios in

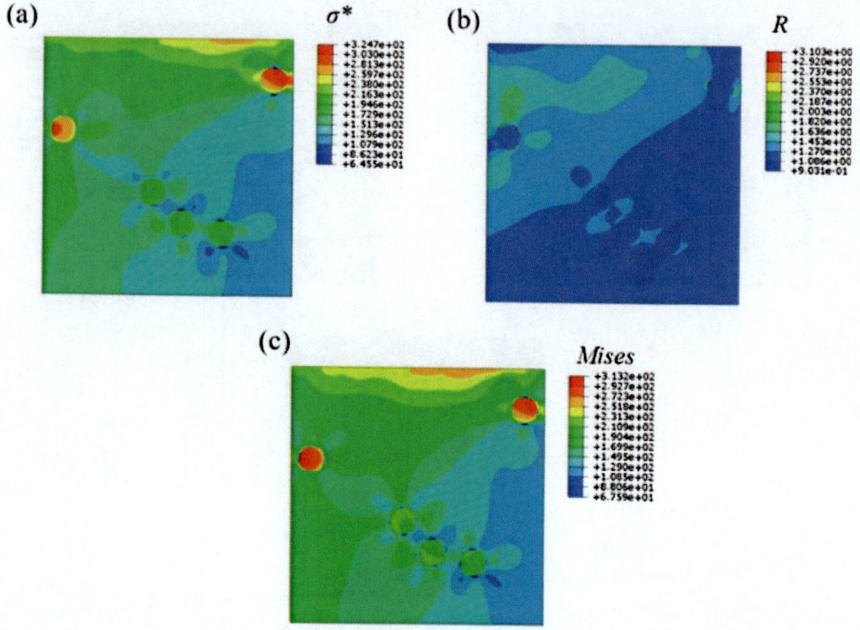

Fig. 6 Distribution of damage parameters and stress of models containing inclusions (Al_2O_3): **a** equivalent multiaxial stress, **b** triaxiality function, and **c** Von Mises stress

each case of micro-void size, it can be noticed that the life ratios are between 0.6 and 0.7, the difference is about 0.1. Therefore, compared with the effect of micro-voids sizes on crack initiation lifetime, the distributions of micro-voids play a more important role.

A similar phenomenon can be noticed in Fig. 7b and c. When the size of inclusions is the same, the difference in lifetime ratio is largely due to the different distribution of inclusions. When the size of Al_2O_3 is 28.8 μm, the largest difference appears, about 0.27. For Al_2CuMg, the biggest discrepancy is 0.25 when the size of it is 12.6 μm. Compared to the results of micro-voids, it can be seen that the maximum fluctuation of life ratios due to the different distributions appears in the models containing micro-voids.

Figure 7d shows the average life ratio of different defect sizes, it can be seen that the existence of micro-voids significantly reduces the crack initiation life, and the life ratio remains the lowest value in the case of three different sizes. Meanwhile, there is not much difference between the two types of inclusions, that is, the types of inclusions have little effect on the crack initiation life as the size of inclusions changes.

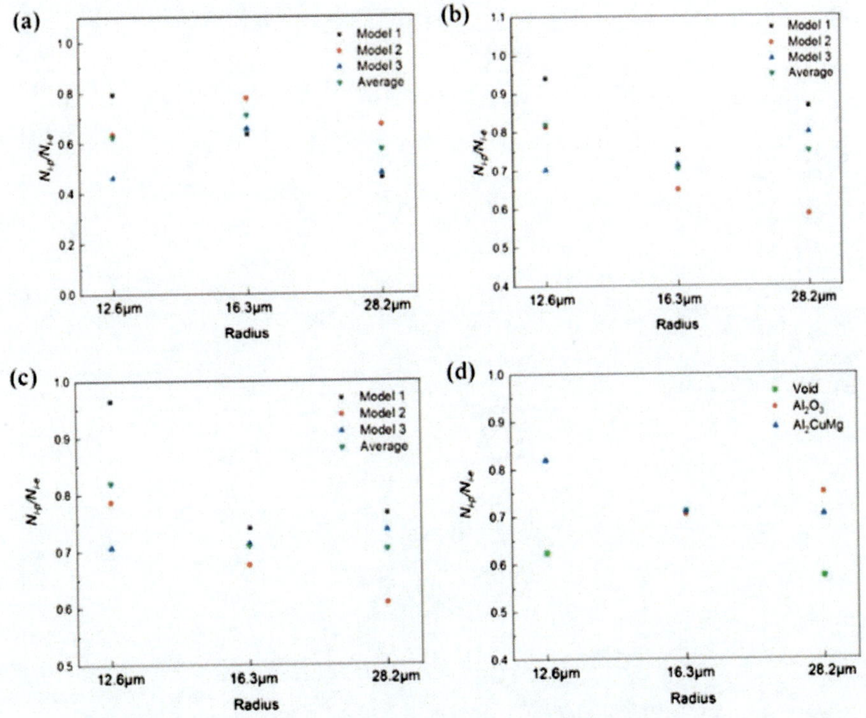

Fig. 7 Life ratio ($N_{i,\,p}/N_{i,\,e}$) in different sizes of defects: **a** micro-void, **b** Al_2O_3, **c** Al_2CuMg and **d** average life ratio comparison among micro-voids and two kinds of inclusions

5.3 *Effect of Defects Volume Fraction on Fretting Fatigue Lifetime*

From Fig. 8a, it can be seen that the locations of micro-voids affect the crack initiation lifetime seriously, the fluctuations of life ratios are 0.55, 0.65, and 0.68 when the volume fraction of micro-voids are 2.04, 4.08, and 6.12%. However, the average life ratios are all about 0.45 in the three volume fractions, the difference is small.

The locations of inclusions play an important role in influencing crack initiation lifetime as well, there is a large difference in the three models (Model Top, Model Bot, and Model Mid) at the same volume fraction of inclusions. Meanwhile, the differences are becoming larger as the volume fraction of inclusions increases, the maximum differences of Al_2O_3 and Al_2CuMg are 0.18 and 0.11, respectively. It should be noticed that the life ratio differences resulting from the location discrepancy of inclusions are less than that of micro-voids.

Figure 8d shows the comparison among micro-voids and inclusions' average life ratio in three volume fractions. The lowest values are from micro-void results, and the difference between the two kinds of inclusion increases as the volume fraction of inclusions goes up. It is known that the elastic modulus of Al_2O_3 is more than

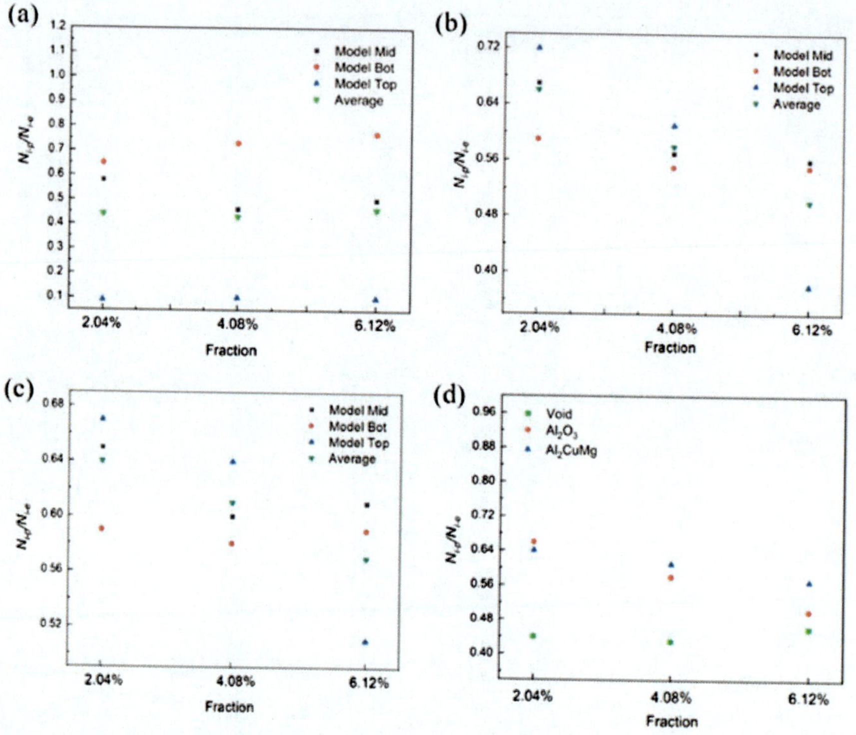

Fig. 8 Life ratio ($N_{i,p}/N_{i,e}$) in different volume fractions of defects: **a** micro-void, **b** Al_2O_3, **c** Al_2CuMg and **d** average life ratio comparison among micro-voids and two kinds of inclusions

three times higher than that of Al_2CuMg (see Table 2), so it can be concluded that inclusions with high elastic modulus reduce the crack initiation life more severely as the volume fraction increases.

6 Conclusions

In this paper, the authors investigated the influence of the characteristics of inclusions and micro-voids on fretting fatigue initiation lifetime. The conclusions are listed below:

1. At the same value of defects' size, the distributions of micro-voids and inclusion affect the crack initiation lifetime dramatically. The size of defects influences the lifetime little.
2. At the same value of size, micro-voids decrease the crack initiation lifetime more than that of inclusions. Meanwhile, there is no obvious discrepancy between the micromodels containing different inclusions.

3. There is not much difference in the results from the different volume fractions of micro-voids, the life ratios are about 0.45.
4. The difference in life ratios between the two kinds of inclusion increases as the volume fraction of inclusions goes up.
5. The effect of micro-voids on lifetime is greater than that caused by inclusions. The inclusions with high elastic modulus reduce the crack initiation life more severely as the volume fraction increases.

Acknowledgements The authors would like to acknowledge the financial support of the grants from the China Scholarship Council (202008130124).

References

1. Nishioka K, Hirakawa K (1969) Fundamental investigations of fretting fatigue: part 4, the effect of mean stress. Bull JSME 12(51):408–414. https://doi.org/10.1299/jsme1958.12.408
2. Jeung H-K, Kwon J-D et al (2015) Crack initiation and propagation under fretting fatigue of Inconel 600 alloy. J Mech Sci Technol 29:5241–5244. https://doi.org/10.1007/s12206-015-1124-8
3. Szolwinski MP, Farris TN (1998) Observation, analysis and prediction of fretting fatigue in 2024-T351 aluminum alloy. Wear 221:24–36
4. Bhatt Nadeem A, Abdel Wahab M (2018) Fretting fatigue crack nucleation: a review. Tribol Int 121:121–138. https://doi.org/10.1016/j.triboint.2018.01.029
5. Kachanov L (1958) Time of the rupture process under creep conditions. Izu Akad Nauk SSR Otd Tech 26–31
6. Chaboche JL (1974) Une loi différentielle d'endommagement de fatigue avec cumulation non linéaire. Office Nationale d'Etudes et de Researches Aérospatiales
7. Lemaitre J (1972) Evaluation of dissipation and damage in metals submitted to dynamic loading. Mech Behav Mater 540–549
8. Lemaitre J, Chaboche J (1975) A non-linear model of creep-fatigue damage cumulation and interaction (for hot metallic structures). In: Mechanics of visco-elastic media and bodies
9. Baidurya Bhattacharya BE (1998) Continuum damage mechanics analysis of fatigue crack initiation. Int J Fatigue 20:631–639. https://doi.org/10.1016/s0142-1123(98)00032-2
10. Quraishi SM, Khonsari MM, Baek DK (2005) A thermodynamic approach for predicting fretting fatigue life. Tribol Lett 19:169–175. https://doi.org/10.1007/s11249-005-6143-7
11. Hojjati-Talemi R, Abdel Wahab M et al (2014) Prediction of fretting fatigue crack initiation and propagation lifetime for cylindrical contact configuration. Tribol Int 76:73–91. https://doi.org/10.1016/j.triboint.2014.02.017
12. Liu ZR et al (2011) The structure and the properties of S-phase in AlCuMg alloys. Acta Mater 59:7396–7405. https://doi.org/10.1016/j.actamat.2011.08.009
13. Prasannavenkatesan R, Przybyla CP et al (2011) Simulated extreme value fatigue sensitivity to inclusions and pores in martensitic gear steels. Eng Fract Mech 78:1140–1155. https://doi.org/10.1016/j.engfracmech.2011.01.027
14. Bhatti Nadeem A, Pereira K, Abdel Wahab M (2018) A continuum damage mechanics approach for fretting fatigue under out of phase loading. Tribol Int 117:39–51. https://doi.org/10.1016/j.triboint.2017.08.009
15. Hojjati-Talemi R, Abdel WM (2013) Fretting fatigue crack initiation lifetime predictor tool: using damage mechanics approach. Tribol Int 60:176–186. https://doi.org/10.1016/j.triboint.2012.10.028

16. Can W, Wenbin Z, Chao L, Magd Abdel W (2022) Two-scale analysis of fretting fatigue in heterogeneous materials with continuum damage mechanics approach. Int J Solids Struct Under Rev
17. Pereira K, Abdel WM (2017) Fretting fatigue crack propagation lifetime prediction in cylindrical contact using an extended MTS criterion for non-proportional loading. Tribol Int 115:525–534. https://doi.org/10.1016/j.triboint.2017.06.026

Wear

Effects of Load and SiC_p Addition on Wear Behaviour of Powder Injection Moulded Aluminium Composite

Tapany Patcharawit, Arada Ngeekoh, and Nutthita Chuankrerkkul

Abstract This research investigated wear behaviour of powder injection moulded SiC_p—reinforced aluminium composite, utilized in applications in small and complex electrical/mechanical parts. Effects of load and SiC_p addition on friction coefficient, specific wear rate and wear mechanisms were investigated. Injection moulding of 5–20 vol.% SiC_p—aluminium feedstocks, prepared at 55% solid loading, was carried out prior to sintering at 645–660 °C/2 h followed by precipitation-hardening at 160 °C/6 h. Effects of loading over 5–65 N range was investigated following the ball—on—flat wear test without lubrication. It was found that SiCp addition promoted wear resistance, providing the lowest specific wear rate at 9.6×10^{-5} mm3/Nm and the friction coefficient of 0.04. Abrasive wear was found to dominate at higher SiCp content. Increasing load resulted in greater wear severity, resulting in greater material loss. Wear test towards high load and high SiCp addition caused mutual materials transfer and the formation of mechanically mixed layer (MML), being composed of crushed SiCp in the hardened and re-oxidized aluminium matrix via comminution under reciprocal sliding wear action. Subsequent fatigue crack resulted in delamination of the MML causing wear debris. In addition, fatigue wear could be prolonged, which was attributable to precipitation hardening of the aluminium matrix.

Keywords Aluminium composite · Silicon carbide · Abrasive wear · Adhesive wear · Mechanically mixed layer · Powder injection moulding

T. Patcharawit (✉) · A. Ngeekoh
School of Metallurgical Engineering, Institute of Engineering, Suranaree University of Technology, 111 University Avenue, Muang, Nakhon Ratchasima 30000, Thailand
e-mail: tapany@sut.ac.th

N. Chuankrerkkul
Metallurgy and Materials Science Research Institute, Chulalongkorn University, Phayathai Road, Wangmai, Pathumwan, Bangkok 10330, Thailand

M. Abdel Wahab (ed.), *Proceedings of the 10th International Conference on Fracture Fatigue and Wear*, Lecture Notes in Mechanical Engineering,
https://doi.org/10.1007/978-981-19-7808-1_10

1 Introduction

Significant growth of the global powder market renders greater opportunities for powder metallurgy (PM) to extend its applications not only limited to automotive, aerospace, dental and medical, but also in the field of electronics, industrial, jewellery, consumer, and etc. Powder injection moulding (PIM) shares similarity to that of PM merits in terms of formulation flexibility, material utilization, high properties, precision surface finish, low cost, and especially focusing mainly on small and complex geometry at high production volume. The process attributes allow particulate composite production for example in the case of SiC_p reinforced aluminium composites where the key steps involve feedstock preparation, injection moulding, debinding, and sintering. The metal powders and reinforcement are well mixed to produce homogeneous feedstock ready for injection moulding. A decent flow of feedstock into the mould cavity is controlled by rheology of the binder, powder morphology and % solid loading to yield sound green sample. Solvent debinding selectively removes soluble polymeric binder allowing open pore structure to facilitate subsequent binder burnout. Sintering should be done in a specific atmosphere over appropriate temperature and time to promote consolidation and densification of the whole structure. Liquid phase sintering permits full densification as the transient liquid fill in the remaining pores in the structure, with the expense of shrinkage and distortion. Theoretical density of 96–98% is well accepted for full densification that is related to remarkably good properties such as hardness, strength, toughness and corrosion as well as wear resistance.

Particulate reinforced aluminium composites are usually produced by stir casting, liquid infiltration, semi-solid casting and powder metallurgy, which mostly aim for property improvement to avoid mechanical failure such as wear and fatigue during services. Dominate wear mechanism of discontinuously reinforced aluminium composite for automotive engine components such as cylinder blocks, pistons, and piston insert rings was indicated to be adhesive wear [1]. The controlling parameters are applied load, sliding speed, wear surface hardness, reinforcement fracture toughness and morphology. Since aluminium is inherently softer than its counterparts such as steels or stainless steels, reinforcement is believed to give enhanced wear resistance. Roles of reinforcement on wear behaviours of the composites are therefore of key interest. In the cast composites, wear rate was found to decrease with increasing SiC_p volume fraction [2–4] due to induced plastic constrain and improved high temperature strength/hardness of the matrix. Effect of SiC_p addition on coefficient of friction on the other hand showed controversially results, which might be due to the reinforcement size along with its distribution and cohesion to the matrix. Lower friction coefficient was due to particle protrusion helped protecting the softer matrix from intimate or severe contacts to counterpart materials. It was signified that 10–20 vol.% SiCp reinforcement in 7xxx and 2xxx series reduced the wear rate, increased seizure temperature but increased friction coefficient [5].

SiC_p reinforcement was also observed to change the dominate wear mechanisms or prolong severe wear regime of the composite in comparison to those of the virgin alloys. Dry sliding wear test in 20 vol.% SiC_p-356Al, showed three wear rate regimes where the effect of SiC_p reinforcement found to be dependent on the applied load [6]. In the first regime (< 10 N) of wear, the composite gave lower wear rate than that of unreinforced alloys and both exhibit mutual materials transfer on the worn surfaces by the action of SiC_p. The formation of α-Fe_2O_3 layer on the composite surface then improved wear resistance. With increasing load (10–95 N) in the intermediate regime of wear, both composite and unreinforced alloys showed similar wear rates where subsurface delamination was observed. Cracking of SiC_p on the composite surface due to higher loads seemed not to give any advantageous over the softer matrix with the Si phase. In the final wear regime (> 98 N), severe adhesive wear was dominated in which the Al matrix was heavily deformed at approximately 250 μm. The inferior wear resistance in the unreinforced alloy was attributed to work softening, which gave massive deformation. SiC_p reinforcement was therefore found to inhibit transition of wear into the severe wear by providing thermal stability to the matrix. In stir cast AA6061-T6/AlNp composite under dry sliding wear, it was indicated that greater load resulted in larger contact area and induced more plastic flow in the matrix by frictional heat and shear stress developed during wearing [7]. The dominate wear mechanism then changed from adhesive wear in the based alloy to abrasive wear in the composite.

Heterogeneous tribo-layer was observed to form over the worn surfaces during the wear, causing two effects: (i) acting as a lubricant layer and (ii) acting as a source of wear debris [8] Most wear studies in aluminium composite have focused mainly on parts produced from liquid or semi-solid state production with small area being conventional powder metallurgy. However, wear properties and mechanisms of the composites produced by powder injection moulding aiming for the production of small engineering parts such as mini gears and electrical transmision has not been widely investigated. Therefore, this research studied wear properties and mechanisms of 5–20 vol.% SiC_p—reinforced aluminium composite produced by powder injection moulding. This aimed to construct the composite wear maps based on specific wear rate and friction coefficient and to identify wear mechanisms of the composite in the as-sintered and precipitation-hardened conditions.

2 Experimental

2.1 Composite Fabrication

Commercial Al-Cu alloy powder of 4.77% Cu, 1.11% Mg, 1.01% Si, and 0.12% Fe composition by weight with an average size of 73 μm size was used as the matrix.

Mixing of SiC_p having 20.0 μm average size at 5, 10, 15, and 20 vol.% to the aluminium matrix was carried out in a ball mill prior to feedstock preparation at 55% solid loading using a multicomponent binder. Powder injection moulding was operated at 170 °C, 90 °C mould temperature and 45 MPa moulding pressure. Solvent debinding was carried out in hexane prior to sintering in a nitrogen atmosphere at 645–660 °C/2 h respective to SiC_p addition to give effective liquid phase sintering. Precipitation hardening was carried out at 160 °C/6 h in order to compare wear test result to that of as-sintered one.

2.2 *Composite Characterization*

The composite in the as-sintered and precipitated hardened conditions was subjected to microstructure and phase analysis using optical microscope, SEM coupled with EDS, and XRD techniques. Sintered bulk density was measured and macro Vickers hardness was carried out.

2.3 *Ball-on-Flat Wear Test*

Following the ASTM G133-95, A Bruker UTM tribolab was employed for ball-on-flat wear test over load varying at 5, 25, 45, and 65 N, a reciprocal length of 10 mm over a 100 m test distance, at frequency of 5 Hz, with no lubrication. The mating material is a 3/16" diameter stainless steel ball having 62 HRC. Real-time friction coefficient was obtained according to the measured normal and tangential forces, F_z and F_x, respectively. Wear maps were constructed based on friction coefficient and specific wear rate. Wear weight loss (W_L) were measured to achieve wear volume loss (V_L) and specific wear rate (W_s) according to Eqs. (1) and (2), respectively, where ρ sintered is the density of the composite and s is the wear distance. The LEXT 3D measuring laser microscope OLS 4100 was used to investigate cross section of the wear groove and wear volume loss in some conditions. Wear surfaces and debris were investigated under SEM JEOL 6010LV coupled with EDS analyser in order to reveal wear mechanisms.

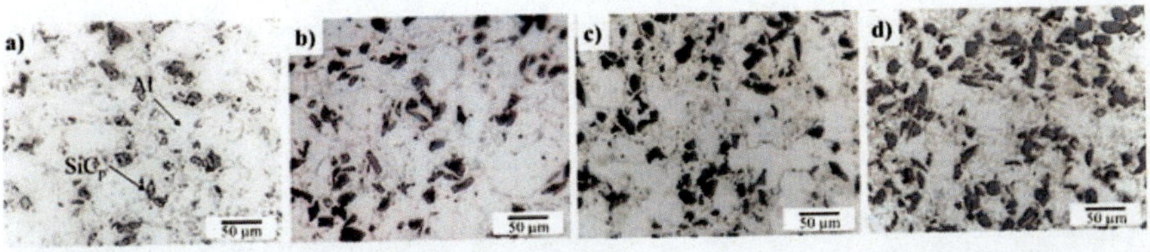

Fig. 1 Optical micrograph of aluminium composite at **a** 5 vol.%, **b** 10 vol.%, **c**, 15 vol.%, **d** 20 vol.% SiC_p addition in the as-sintered condition

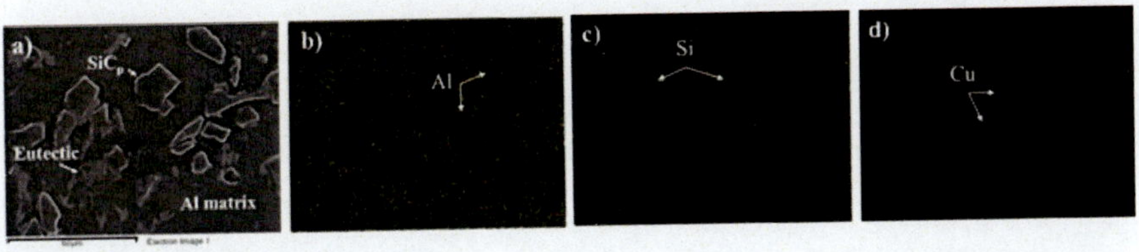

Fig. 2 SEM micrograph and EDS map analysis of 20 vol.% SiC_p reinforced aluminium composite in the as-sintered condition

3 Experimental Results

3.1 Microstructure and Phase Analysis of Al-SiC_p Composite

Optical microstructure of as-sintered composite in Fig. 1 showed angular SiC_p uniformly embedded in the aluminium matrix. Clustering of SiC_p with micro-porosity was evident at SiC_p addition of higher than 15 vol.%. Microstructure and phase analysis by SEM coupled with EDS mapping and point analysis in Figs. 2 and 3 revealed good interfacial bonding between SiC_p and aluminium matrix, indicating effective liquid phase sintering. EDS point analysis suggested phases appearing as eutectic (spectrum 2), Al2Cu (spectrum 4), and Mg_2Si (spectrum 5), corresponding to the aluminium alloy composition. Moreover, XRD analysis as demonstrated in Fig. 4 indicated the existence of the phases previously mentioned along with aluminium nitride (AlN) as a result of liquid phase sintering in nitrogen atmosphere via direct mechanism [9]. Composite in the precipitation hardened condition showed similar phases and microstructure.

3.2 Sintered Density and Macro Vickers Hardness

Sintered density of the composite are in the range 2.43–2.68 g/cm^3 as listed in Table 1. The optimum SiC_p addition is at 10–15 vol.%. If the densities of Al-Cu

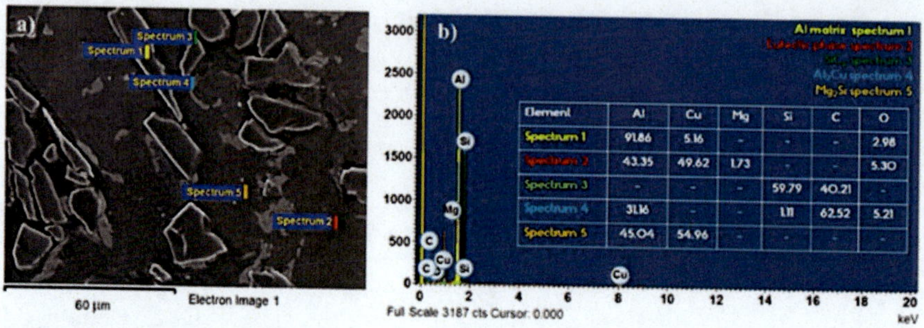

Fig. 3 SEM micrograph and EDS point analysis of 20 vol.% SiC_p reinforced aluminium composite in the as-sintered condition

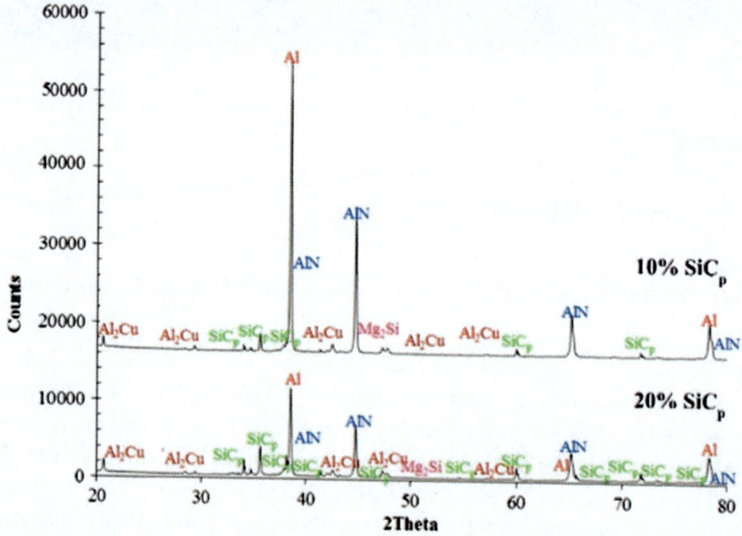

Fig. 4 XRD analysis of as- sintered 10 and 20 vol.% SiC_p reinforced aluminium composite

alloy are taken as 2.78 [10] and 3.20 g/cm^3, respectively, % theoretical density can be obtained up to the value of 94.27%, by considering micro-porosity observed. Although the composite has been subjected to liquid phase sintering, greater amount of SiC_p content produced SiC_p clustering and retarded liquid flow and solid grain rearrangement to allow proper pore filling along SiC_p–SiC_p interfaces. The macro Vickers hardness values provided similar trends where precipitation hardening offered significantly higher hardness values, giving the maximum of 166.6 Hv.

Table 1 Physical and mechanical properties of 5–20 vol.% SiC_p—reinforced aluminium composite fabricated by powder injection moulding

SiC_p addition (%)	Sintered density (g/cm^3)	% Theoretical density	Macro vickers hardness (H_v)	
			As-sintered	Precipitation hardened
5 vol.	2.47	88.18	123.9	132.9
10 vol.	2.63	93.20	132.4	160.8
15 vol.	2.68	94.27	138.9	166.6
20 vol.	2.43	84.85	125.0	164.2

3.3 *Wear Properties*

Wear Groove Width. After the wear test, the wear groove width increased with increasing load from 5 to 65 N, whereas the addition of SiC_p reduced the wear track width as comparatively shown in Fig. 5a–h. The wear track widths of as-sintered and age-hardened specimens were quite comparable. The minimum–maximum values were measured to be 0.17–2.22 mm. Cross sections of wear grooves of 10 vol.% SiCp-Al composite in the as-sintered condition are illustrated in Fig. 6a–d. It is shown that testing at the greater load of 45 N in comparison to 25 N gave relatively wider wear groove with ragged wear groove profile and rougher surface.

Coefficient of Friction. Real time normal and tangential forces (F_z and F_x) along with coefficient of friction (COF) are plotted over selected time to 400 s as shown in Fig. 7a–d, whereas the COF plots against load and COF maps are illustrated in Figs. 8 and 9, respectively. It is observed that the COF of 5–20 vol.% SiCp reinforced aluminium composite is 0.04–0.76, depending on SiC_p addition. It can be seen that the COF of 10 vol.% SiC_p composite was the lowest whereas the highest COF was found in 20 vol.% SiCp composite. Variation of loads at 20 vol.% SiC_p composite is rather high, as shown in Fig. 7d. It might possibly be that increasing SiC_p content would increase area of surface contact between SiC_p and mating material, hence increasing the friction during wear action. COF maps were constructed as exhibited in Fig. 9 to be dependent on load and SiC_p addition for both as-sintered and precipitation hardened conditions, in which the latter appeared to give slightly higher the COF in this case.

Specific Wear Rate. Wear weight loss (W_L) of the sample measured after wear testing was used to calculate the specific wear rate (W_s) through Eqs. (1) and (2);

$$V_L = \left(W_L \times 10^{-3}\right) \cdot \rho_{\text{sintered}} \tag{1}$$

$$W_s = V_L/(F_z \cdot s) \tag{2}$$

where W_L is the wear weight loss, g

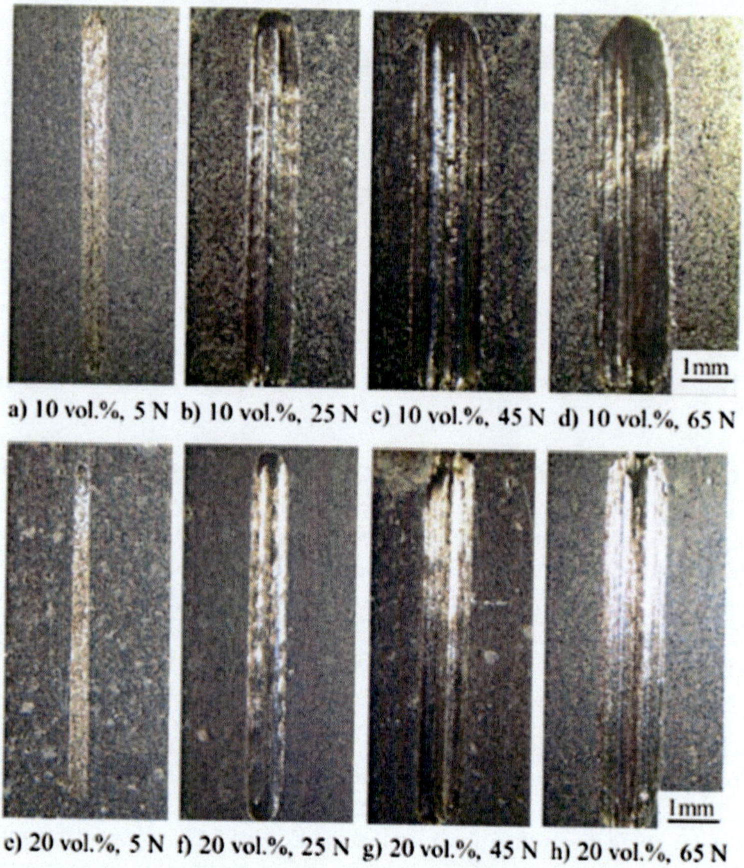

Fig. 5 Wear groove widths of **a–d** Al-10 vol. % SiC_p composite and **e–h** Al-20 vol. % SiC_p composite, both with increasing load from 5 to 65 N

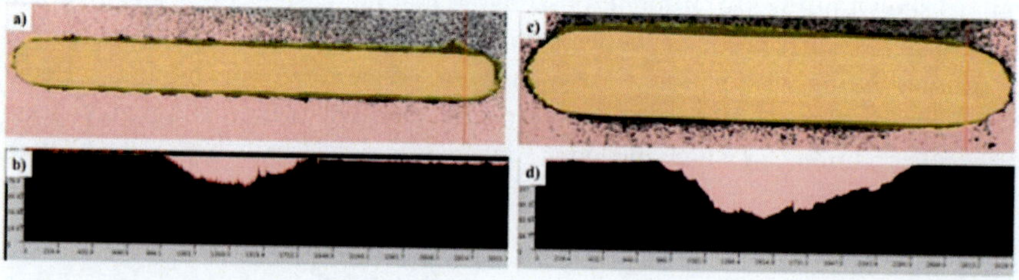

Fig. 6 Cross section profiles of wear groove width for as-sintered 10 vol.% SiC_p composite tested under **a** and **b** 25 N load, **c** and **d** 45 N load

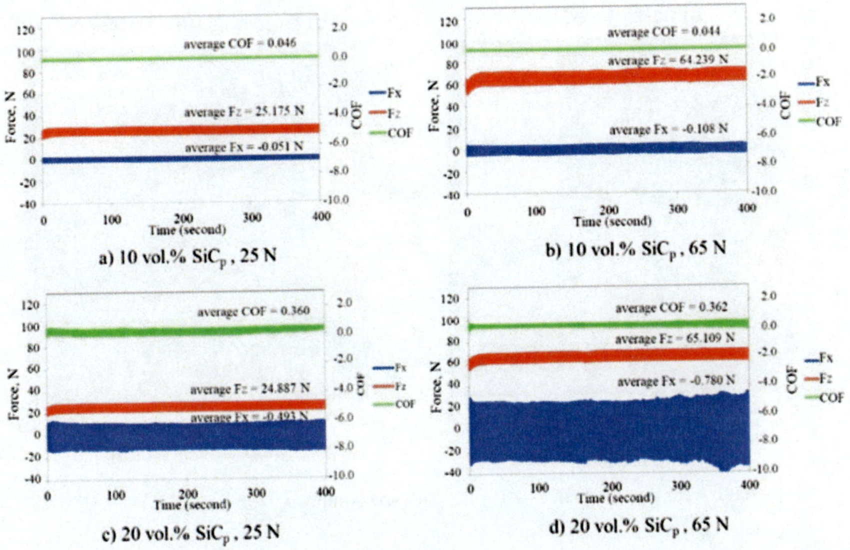

Fig. 7 COF, F_x, and F_z plots against time for 10 and 20 vol.% SiC_p reinforced aluminium composite at 25 and 65 N in precipitation hardened condition

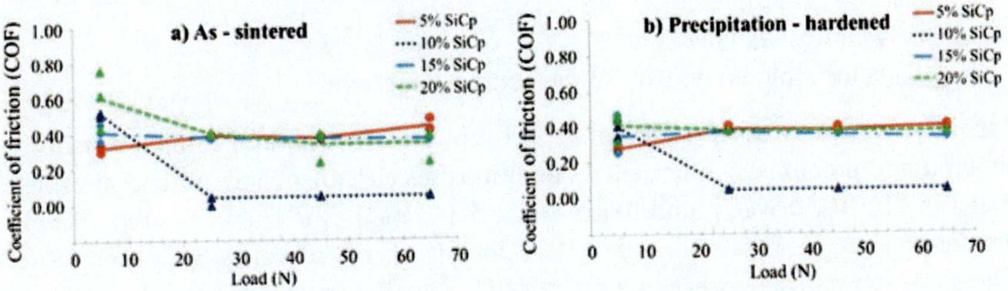

Fig. 8 COF of 5–20 vol.% SiC_p reinforced aluminium composite in **a** as-sintered and **b** precipitation hardened conditions

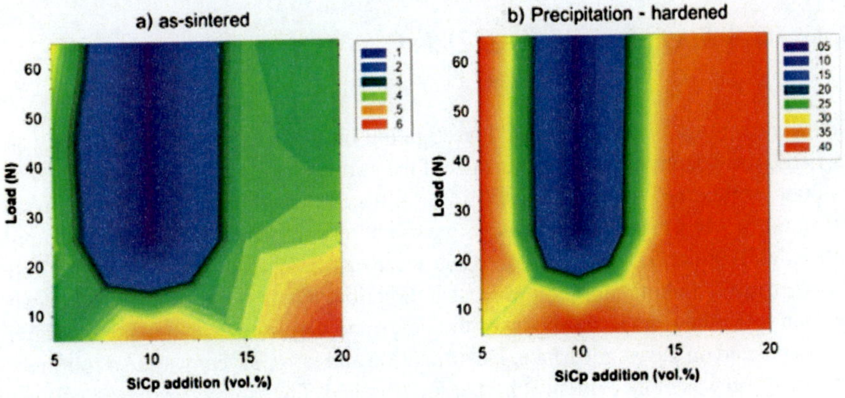

Fig. 9 COF maps of 5–20 vol.% SiC_p-Al composite

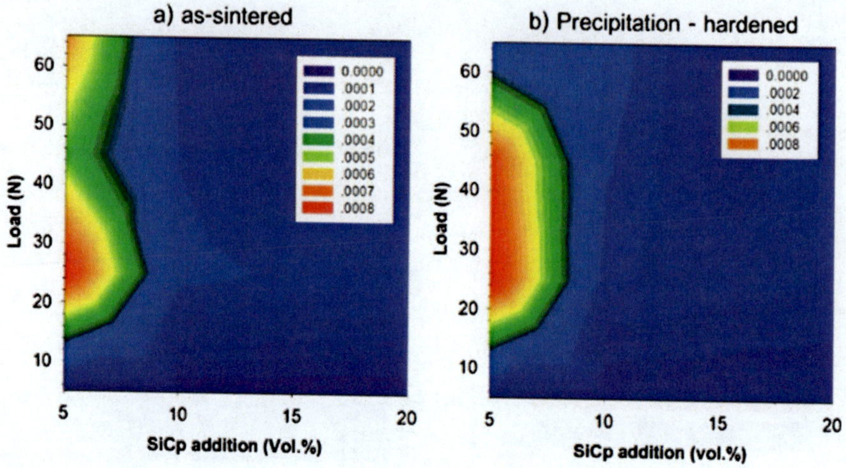

Fig. 10 Specific wear rate maps of 5–20 vol.% SiC_p-Al composite. (unit: mm^3/Nm)

V_L is the wear volume loss, mm^3
W_s is the specific wear rate, mm^3/Nm
F_z is the vertical force, N
s is the wear test distance, m
$\rho_{sintered}$ is the sintered density of the composite, g/mm^3

Specific wear rate maps of 5–20 vol.% SiC_preinforced aluminium composite in as—sintered and precipitation hardened conditions tested under 5–65 N load are illustrated in Fig. 10. It was found that 5 vol.% SiC_p composite showed comparatively greater specific wear rate than those obtained from 10–20 vol.% SiC_p composite. The specific wear rates were in a range of 10^{-3} to 10^{-5} mm^3/Nm and the effect of SiC_p addition is evidently seen in this case.

3.4 *Wear Surfaces and Wear Debris*

Investigation on Wear Surfaces. Investigation on wear surfaces under SEM was carried out to identify wear mechanisms of the composite after ball-on-flat wear test without lubrication. According to Fig. 11a–f, the dominant wear mechanisms were observed to be abrasive and adhesive wear, and accompanied by fatigue, delamination, and oxidation wear. The abrasive wear can be characterized as macro and micro grooves running along the wear test direction, whereas the adhesive wear can be seen from the transferred layer of the deformed aluminium matrix layer, which was deposited on the wear surface. Delamination and fatigue wear was also routinely observed and was more evident at higher load applied. The fatigue wear can be found more severe with increasing load and well characterized by parallel cracks running

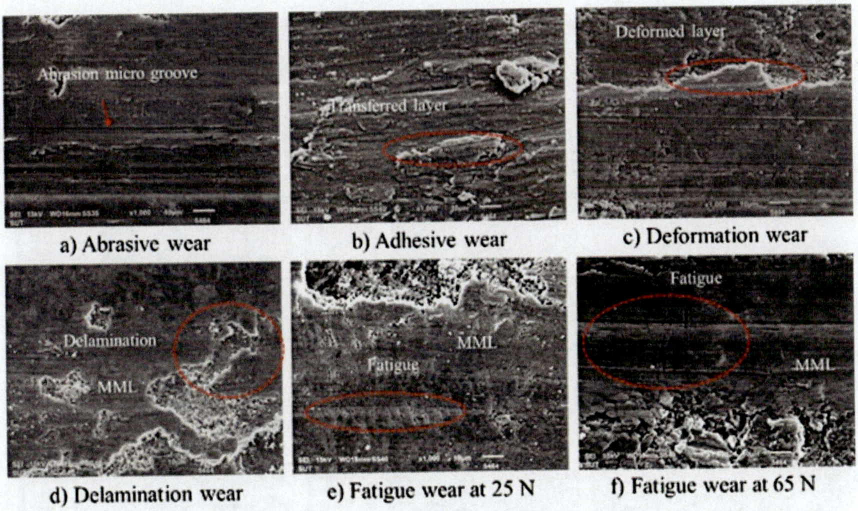

Fig. 11 General wear types of SiC_p aluminium composite subjected to ball-on-flat wear test

perpendicular to the test direction. It should be noted that precipitation hardened specimens showed the fatigue wear to appear at a later stage in comparison to that of as-sintered specimens under the same wear test condition. Increasing load resulted in increasing severity of wear.

EDS point and area analysis of wear surface tested at 25 N load in Fig. 12 showed the key elements being Al and O along with O, Si, and Mg, resembling those of the matrix. Increasing load to 65 N promoted mutual materials transfer between the mating materials where Fe (30.16 wt.%) and Cr (3.43 wt.%) were detected on wear surface. Hard SiC_p might have scratched on the stainless steel ball counterpart, giving debris transferrable on to the composite surface as exhibited in Fig. 13. Furthermore, greater amount of load also influenced the deformation characteristics of the matrix. The aluminium matrix was deformed and sheared. Repetitive wear cycle correspondingly promoted re-oxidation of wear surface, giving relatively greater O content up to 46 wt.% in the wear debris than that found in the base metal. Greater SiC_p addition seemed to promote abrasive wear, whereas lower SiC_p content allowed adhesive wear to predominate especially at higher load of 65 N. Tables 2 and 3 summarize dominant wear mechanisms under 5–65 N load of 5–20 vol.% SiCp PIMed composite in the as-sintered and precipitation hardened conditions.

Investigation on Wear Debris. The composite wear debris can be categorized as flake and agglomerate types with their sizes becoming larger at higher load, as shown in Fig. 14a, b, comparable to result found in SiC_p reinforced 356Al composite under high load [11]. Evidences showed that hardened surface appeared crack and later be delaminated. Large delaminated debris then became flakes. Agglomerated types might have caused by breaking down of these flake typed wear debris by repeated shear action in the three body abrasion wear. If the SiC_p was finally pulled out and

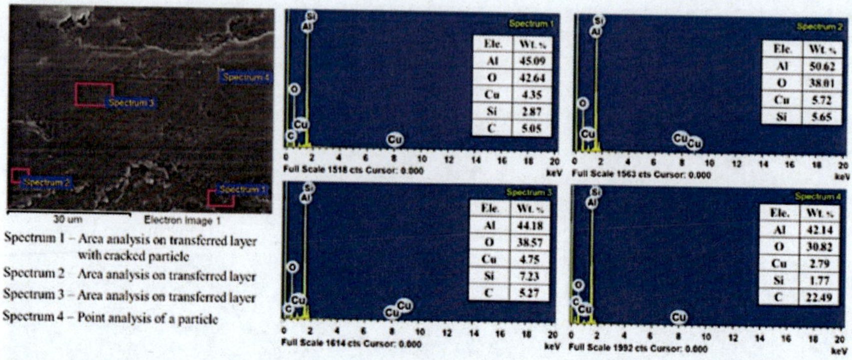

Fig. 12 EDS analysis of precipitation hardened 10 vol.% SiC_p aluminium composite wear surface under 25 N

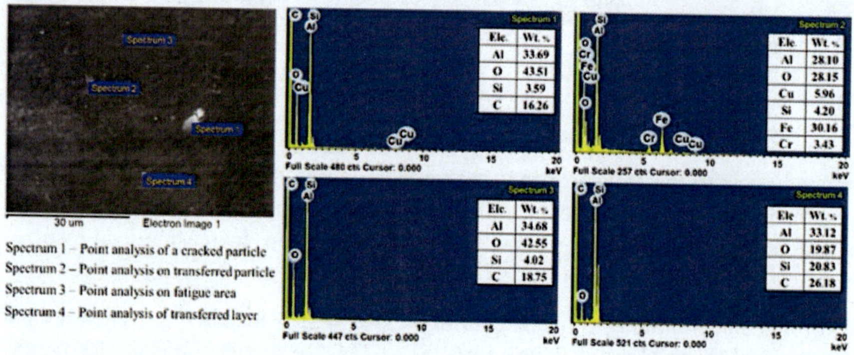

Fig. 13 EDS analysis of precipitation hardened 20 vol.% SiC_p aluminium composite wear surface under 65 N

Table 2 Wear mechanisms of SiC_p reinforced aluminium composite after sintering

	Load = 5 N	Load = 25 N	Load = 45 N	Load = 65 N
5 vol.% SiC_p	Abrasive wear Adhesive wear	Adhesive wear Abrasive wear Fatigue wear	Adhesive wear Abrasive wear Fatigue wear	Adhesive wear Abrasive wear Fatigue wear
10 vol.% SiC_p	Abrasive wear Adhesive wear	Adhesive wear Abrasive wear Fatigue wear	Adhesive wear Abrasive wear Fatigue wear	Adhesive wear Abrasive wear Fatigue wear
15 vol.% SiC_p	Abrasive wear	Adhesive wear Abrasive wear Fatigue wear	Abrasive wear Adhesive wear Fatigue wear	Abrasive wear Adhesive wear Fatigue wear
20 vol.% SiC_p	Abrasive wear	Abrasive wear Adhesive wear Fatigue wear	Abrasive wear Adhesive wear Fatigue wear	Abrasive wear Adhesive wear Fatigue wear

Table 3 Wear mechanisms of SiC_p reinforced aluminium composite after precipitation hardening

	Load = 5 N	Load = 25 N	Load = 45 N	Load = 65 N
5 vol.% SiC_p	Abrasive wear Adhesive wear	Adhesive wear Abrasive wear Fatigue wear	Adhesive wear Abrasive wear Fatigue wear	Adhesive wear Abrasive wear Fatigue wear
10 vol.% SiC_p	Abrasive wear Adhesive wear	Adhesive wear Abrasive wear Fatigue wear	Adhesive wear Abrasive wear Fatigue wear	Adhesive wear Abrasive wear Fatigue wear
15 vol.% SiC_p	Abrasive wear	Adhesive wear Abrasive wear Fatigue wear	Abrasive wear Adhesive wear Fatigue wear	Abrasive wear Adhesive wear Fatigue wear
20 vol.% SiC_p	Abrasive wear	Abrasive wear Adhesive wear Fatigue wear	Abrasive wear Adhesive wear Fatigue wear	Abrasive wear Adhesive wear Fatigue wear

its size was considered to be larger than the gap between the mating materials at the contact, parallel furrow or plowing might occurred under no lubrication condition. Figure 14d revealed multiple sheared layer as the cause of repetitive wear motion to produce plowing of deformable Al matrix by SiC_p. The plow groove size was observed to be similar to the size of the SiC_p that might have been crushed.

Wear Mechanisms and Sequence. Following extensive wear surfaces and wear debris investigation, the wear mechanisms of the SiC_pAl composite fabricated by

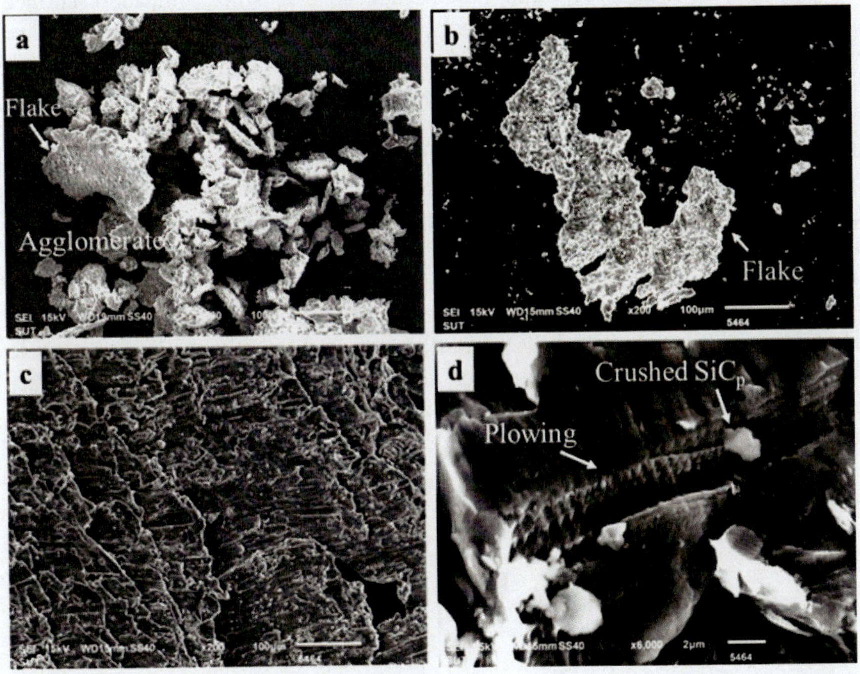

Fig. 14 Flake and agglomeration types of wear debris

powder injection moulding under unlubricated ball-on-flat wear test can be determined. In the as-sintered condition, at low SiC_p addition of 5–10 vol.% and low load (5 N), the predominated wear was found to be abrasive and became adhesive as the load increased to 25–65 N for both as-sintered and precipitation-hardened conditions. By increasing SiC_p addition to 15 and 20 vol.%, abrasive wear became dominant. In the precipitation hardened condition, the composite showed quite similar wear behaviour to that of the as-sintered condition, except for the fatigue wear has been prolonged at higher load. It might be that finer precipitates of Al_2Cu, Mg_2Si, and AlN facilitated strain-harden effect to the matrix that helped to retard fatigue crack initiation and void nucleation. Behaviours of fatigue and delamination wear can be explained that the transferred layers of the hardened and re-oxidized aluminium alloy matrix with cracked SiC_p were repeatedly comminuted, leading to the so-called mechanically mixed layer (MML) formation [12]. Fatigue cracks then originated in the MML due to repetitive cycles of sliding wear action and finally delayered.

The wear sequence of SiC_p reinforced aluminium fabricated by powder injection moulding as illustrated in Fig. 15 involved matrix plastic deformation and flow in the first stage. Fracture of SiC_p reinforcement occurred along with mutual materials transfer. Repeated motion of sliding wear resulted in the MML formation, through hardening of the aluminium matrix and surface oxidation. The compaction zone included the hardened and re-oxidized MML and the based SiC_p aluminium composite on the top surface. Comminution then led to densified compaction surface and subsequently produced fatigue crack. MML delamination took place and gave wear debris in the flake and agglomeration forms.

According to the experimental results so far, it has been shown that aluminium composite having 10 vol.% SiCp addition had the lowest COF and generally showed the transformation of the wear mechanism from adhesive to abrasive towards higher load level. It might be that the MML formation during wear action helped to reduce COF and the specific wear rate [13], giving better wear resistance performance in 10 vol.% SiCp aluminium composite. Furthermore, 10–15 vol.% SiC_p addition gave optimum sintered density and hardness values, due to uniform distribution of SiC_p in the aluminium matrix with least degree of SiC_p clustering and micro-porosity. Therefore, smaller amount of SiC_p addition at 5 vol.% might not be sufficient to reduce mating contact area and provided less hardening effect to the matrix. Higher SiC_p addition of 15–20 vol.% was prone to abrasive wear which promoted three body wear and severity.

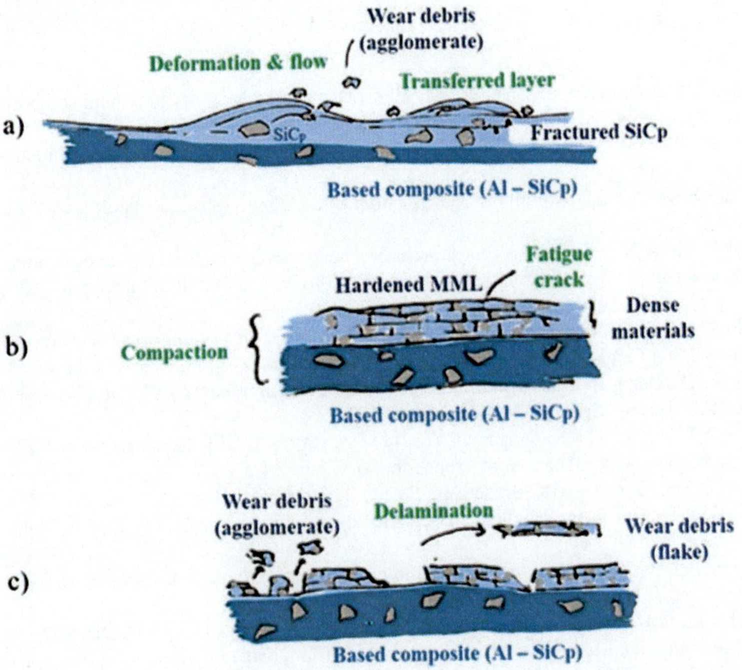

Fig. 15 Wear sequence of SiC_p reinforced aluminium under ball-on-flat wear test

4 Conclusions

According to experimental results, conclusions can be drawn as follows;

1. The aluminium composite reinforced with 5–20 vol.% SiC_p addition possessed the specific wear rate of 10^{-3}–10^{-5} mm^3/Nm and COF of 0.04–0.76 over 5–65 N load range under sliding wear without lubrication. Aluminium composite reinforced with 10 vol.% SiC_p provided optimum wear performance.
2. The dominant wear mechanisms were abrasive and adhesive along with fatigue and oxidation, where severity was raised by increasing load.
3. Transformation of wear mechanisms is due to increasing SiC_p and load. Greater SiC_p addition of 15–20 vol.% promoted wear resistance and wear mechanism became abrasive. Increasing load facilitated plastic deformation of the matrix to promote adhesive wear and mutual material transfer.
4. The MML formation was through comminution of hardened and oxidized aluminium matrix with fractured SiC_p due to reciprocal sliding wear.
5. Precipitation hardening slightly increased the COF and helped to prolong fatigue wear, by resisting fatigue crack and void nucleation.

References

1. Deuis RL, Subramanian C, Yellup JM (1997) Dry sliding wear of aluminium composites—a review. Compos Sci Technol 57(4):415–435
2. Rohatgi PK (1988) Cast metal matrix composites metal hand book, 9th edn. ASM International, Materials Park
3. Surappa MK (2003) Aluminum matrix composites: challenges and opportunities. Sadhana 28:319-334
4. Gul F, Acilar M (2004) Effect of the reinforcement volume fraction on the dry sliding wear behaviour of Al–10Si/SiCp composites produced by vacuum infiltration technique. Compos Sci Technol 64, 1959–1970
5. Rao RN, Das S, Mondal DP, Dixit G, Tulasi DSL (2013) Dry sliding wear maps for AA7010 (Al–Zn–Mg–Cu) aluminium matrix composite. Tribol Int 60:77–82
6. Alpas AT, Zhang J (1992) Wear rate transitions in cast aluminium - silicon alloys reinforced with SiC particle. Scr Metall Mater 26:505–509
7. Kumar BA, Murugan N, Dinaharan I (2014) Dry sliding wear behavior of stir cast AA6061-T6/AlNp composite. Trans Nonferrous Met Soc China 24:2785–2795
8. Uyyuru RK, Surappa MK, Brusethaug S (2006) Effect of reinforcement volume fraction and size distribution on the tribological behavior of Al-composite/brake pad tribocouple. Wear 260:1248–1255
9. Hai NH, Tuan NQ (2016) Analysis on the formation of AlN particles via gas/liquid reaction in-situ. Int J Emerg Technol Adv Eng 6(11):1–8
10. Melúch L (2019) Warm compaction of aluminium alloy alumix 123, Ph.D. thesis, Department of Metallurgy and Materials, The University of Birmingham
11. Li XY, Tandon KN (2000) Microstructural characterization of mechanically mixed layer and wear debris in sliding wear of an Al based composite. Wear 245:148–162
12. German RM (2012) Metal powder injection molding (MIM): key trends and markets. In: Handbook of metal injection molding. Woodhead Publishing, pp 1–25
13. Rajesh AM, Kaleemulla KM, Saleemsab D, Bharath KN (2019) Generation of mechanically mixed layer during wear in hybrid aluminium MMC under as-cast and age hardened conditions. SN Appl Sci